Lecture Notes in Mathematics

Edited by A. Dold and E

Subseries: Department of Mat
University of Maryland, Colleg
Adviser: L. Greenberg

517

Shmuel Glasner

Proximal Flows

Springer-Verlag
Berlin · Heidelberg · New York 1976

Author
Shmuel Glasner
Department of Mathematical Sciences
Tel-Aviv University
Tel-Aviv/Israel

Library of Congress Cataloging in Publication Data

Glasner, Shmuel, 1945-
Proximal flows.

(Lecture notes in mathematics ; 517)
"An elaboration on notes taken ... during a course entitled 'Topics in topical dynamics,' which was given by the author in the spring semester of the academic year 1973-74 at the University of Maryland."
Bibliography: p.
1. Topological dynamics. 2. Lie groups.
3. Harmonic functions. I. Title. II. Series:
Lecture notes in mathematics (Berlin) ; 517.
QA3.L28 no. 517 [QA611.5] 510'.8s [514'.7] 76-9866

AMS Subject Classifications (1970): 43A85, 54H15, 54H20, 54H25, 57E20, 60B05

ISBN 3-540-07689-1 Springer-Verlag Berlin · Heidelberg · New York
ISBN 0-387-07689-1 Springer-Verlag New York · Heidelberg · Berlin

Printing and binding: Beltz Offsetdruck, Hemsbach/Bergstr.

TO

RUTH

PREFACE

This work is an elaboration on notes taken, mainly by Professor N. Markley, during a course entitled "Topics in topological dynamics" which was given by the author in the spring semester of the academic year 1973-74 at the University of Maryland. The main theme on which these notes are based is the notion of proximality. This notion is exploited in two principal directions. The first one is the abstract "algebraic" theory of topological dynamics, created by R. Ellis, and the other is H. Furstenberg's theory of boundaries for Lie groups and of harmonic functions. Admittedly these two theories have different flavors and use different techniques, yet we think that an interaction between them might be fruitful.

A good example of this interaction is Furstenberg's characterization of continuous harmonic functions on a symmetric space $D = \mathbb{G}/\mathbb{K}$ as those continuous functions in $L^{\infty}(D)$ whose orbit closure (in the weak $*$ topology) is a strongly proximal flow, (Theorem VI.3.1.). Other instances are the concrete identification of the universal strongly proximal flow and the generalized strong Bohr compactification of a connected semisimple Lie group with a finite center, (Theorems IV.3.2. and VIII.3.6. respectively).

The notes are divided into ten chapters each starting with a short introduction, which describes the material included and indicates its main sources. The prerequisites for reading the "pure" topological dynamics parts are just point set topology and elementary functional analysis. For the rest a certain knowledge of Lie group theory and of probability is needed. Actually only a very restricted portion of these theories is used; I tried to summarize the necessary results in sections IV.1. and V.3. If one is interested only in the abstract theory of topological dynamics and the theory of PI-flows then he can skip chapter IV, V and VI.

I wish to thank Professors R. Ellis and L. Shapiro with whom I had the pleasure of colaborating on the work "PI-flows" [12]. The major part of Chapters IX and X is based upon this paper. I wish to thank Professor W.A. Veech for several fruitful discussions and for his permission to use his unpublished result (Theorem III.4.1.). My thanks are due to all the participants of the course, and to Professors J. Auslander and R. Lipsman who read parts of the manuscript. Professor H. Furstenberg gave me many helpful suggestions during the writing of the notes which improved the whole work immensely. Finally I would like to warmly thank Professor N. Markley who originated the idea of preparing these lectures notes took notes during the course, corrected many mistakes, simplified proofs and took upon himself the painful task of supervising the printing of the manuscript.

TABLE OF CONTENTS

CHAPTER I
THE UNIVERSAL APPROACH

We give here a brief introduction to the algebraic theory of flows which was developed mainly by R. Ellis. A much better source for this material is the first part of [9]. Nevertheless we present it here for the following reasons. First, to make this notes self contained as much as possible. Second, since we write the group action on the left, refering to [9] might be confusing. Lastly, we use this chapter to introduce our notations which differ from those of [9] mainly in the fact that we use pointed flows rather than algebras.

For more information on the subjects which are discussed in this chapter, the interested reader is referred to the following sources: [22], [8], [1], [2], [16], [39].

I.1 DEFINITIONS

A flow is a triple (T,X,π) where, X is a compact Hausdorff space, T is a topological group and $\pi: T\times X \to X$ is a continuous function satisfying the axioms

(1) $\pi(e,x) = x$ for all $x \in X$, where e is the identity element of T, and

(2) $\pi(ts,x) = \pi(t,\pi(s,x))$ for all $t,s \in T$ and $x \in X$.

We say that T acts on X. Usually we shall write $\pi(t,x) = tx$, unless some confusion may arise as to what is the action of T on X, and we shall refer to the flow (T,X), or X rather than (T,X,π).

Notice that fixing $t \in T$, the map $x \to tx$ is necessarily a homoeomorphism of X onto itself, and the map from t to this homeomorphism is a group homomorphism of T into the group of all homeomorphisms of X onto itself.

Given a flow X, we can look at the family of closed invariant and non-empty subsets of X. Each of these is called a <u>subflow</u> of X. By Zorn's lemma this family contians a minimal element. We say that a flow is <u>minimal</u> if it has no proper subflows. Thus every flow has at least one minimal subflow.

If $x \in X$, the subset $\{tx \mid t \in T\}$ is called the <u>orbit</u> of x and is denoted by $\mathcal{O}(x)$. The orbit closure is denoted by $\bar{\mathcal{O}}(x)$. The flow X is a minimal flow iff $\bar{\mathcal{O}}(x) = X$ for every $x \in X$. A flow X is said to be point transitive if there exists $x \in X$ for which $\bar{\mathcal{O}}(x) = X$.

Given two flows (T,X) and (T,Y) one can form the <u>product flow</u> $(T,X \times Y)$ where $t(x,y) = (tx,ty)$. We say that (T,Y) is a <u>factor</u> of (T,X) or that (T,X) is an extension of (T,Y) if there exists a continuous map $X \xrightarrow{\phi} Y$ onto Y such that $\phi(tx) = t\phi(x)$ for all $x \in X$ and $t \in T$. A continuous map which satisfies this last property is called a homomorphism (of flows). If $X \times Y$ is a product flow then the projection maps are clearly homomorphisms and X and Y are both factors of $X \times Y$.

A general homomorphism need not be onto, but is Y is minimal then ϕ is necessarily onto. X and Y are <u>isomorphic</u> if there exists a ono-to-one homomorphism of X onto Y.

In [16] H. Furstenberg introduced, by analogy with the integers, the notion of disjointness of a pair of flows. We say that X and Y are disjoint if whenever X and Y are factors of some flow Z, then $X \times Y$ is already a factor of Z in such a way that the composition of this map with the projections of $X \times Y$

onto X and Y gives the original maps of Z onto X and Y respectively. For minimal flows X and Y this just means that the flow $X \times Y$ is itself minimal.

One would like to have, as is of course the case with the integers, a theorem about the equivalence of disjointness with having no common factors. It is easy to see that disjointness implies that there exists no non-trivial common factor. However, the other implication is false as was shown by Knapp [30]. Whether this equivalence holds for minimal flows of an abelian group is still an open question.

The points x and y of a flow X are said to be <u>proximal</u> if there exists a net $\{t_i\}$ os elements of T such that $\lim t_i x = \lim t_i y$. The points x and y are <u>distal</u> if either $x = y$ or x and y are not proximal. Thus if x and y are both proximal and distal, they must be equal. We denote

$$P = P(X) = \{(x,y) \in X \times X \mid x \text{ and } y \text{ are proximal}\}.$$

P is called the proximal relation on X. It is clear that P is a symmetric reflexive relation on X and it is an invariant subset of $(T, X \times Y)$. In general. it is not transitive and hence not an equivalence relation. If it is closed then it is transitive [ ]. The converse of this is false [ ].

Let R be a closed invariant equivalence relation on X. The action of T on X induces a flow $(T,X/R)$ on the quotient space and the canonical map π of X into X/R is a homomorphism. So $(T,X/R)$ is a factor of (T,X). Now let ϕ be a homomorphism of (T,X) onto (T,Y) and set $R = \{(x,x') \mid \phi(x) = \phi(x')\}$. Then it is immediale that R is a closed invariant equivalence relation on X and there exists an isomorphism ψ of $(T,X/R)$ onto (Y,Y)

such that $\psi \circ \pi = \phi$. Thus the closed invariant equivalence relations completely determine all the factors of a flow.

We say that the flow (T,X) is *equicontinuous* or *almost periodic* if T acts on X as an equicontinuous family i.e. if for every index α (in the unique uniform structure on X) there exists an index β such that $t\beta \in \alpha$ for all $t \in T$. If X is an equicontinuous flow and $(x,y) \in P(X)$ then $x = y$, because if this is not the case then there is an index α such that $(x,y) \in \alpha$. Now for every index β there is $s \in T$ such that $s(x,y) \in \alpha$. Choose β to be an index such that $t\beta \subset \alpha$ for all $t \in T$, then $(x,y) \in s^{-1}\beta \dot{\subset} \alpha$ is a contradiction. Thus every pair of points is an equicontinuous flow is distal. Flows which satisfy this last property are called *distal flows*. A flow X is called *proximal* if $P(X) = X \times X$.

We demonstrate the above notions with the following two examples.

EXAMPLE A. Let X be the interval $[0,1]$ with identified end points. Let $0 < \alpha < 1$, and define a homeomorphism t of X onto itself, $tx = x + \alpha \pmod 1$. Let T be the group $\{t^n \mid n \in \mathbb{Z}\}$, then (T,X) is an equicontinuous flow (hence also a distal flow). By Kronecker theorem (T,X) is minimal iff α is an irrational number.

EXAMPLE B. Let $X = \{0,1\}^{\mathbb{Z}}$, the space of all bilateral sequences of zeroes and ones. Let t be the shift operator on X, i.e. for $\xi \in X$ let $t\xi$ be defined by $(t\xi)_i = \xi_{i-1}$. Let $T = \{t^n \mid n \in \mathbb{Z}\}$ be the group of homeomorphism of X generated by t.

The following statements are easy to verify.

(1) (T,X) is point transitive but not minimal.

(2) The orbit of $\xi \in X$ is dense iff every block of zeroes and ones appears in ξ.

(3) $\xi,\eta \in X$ are proximal iff they have arbitrarily large common blocks in the same position.

(4) The orbit closure of $\xi \in X$ is minimal iff given a block B either B does not appear in ξ at all or there exists n such that every block of length n in ξ contains a copy of B.

I.2 THE STONE-ČECH COMPACTIFICATION OF THE DISCRETE GROUP T.

Clearly the notions of point-transitivity, minimality, proximality and equicontinuity do not depend on the topology of the acting group. This is also the case with many other notions in topological dynamics. It is therefore worthwhile to consider, given a flow (T,X), the associated flow (T_d,X) where T_d denotes the topological group T made discrete.

For a discrete group T, we shall now describe the algebraic structure of the semigroup βT (the Stone-Čech compactification of T). We shall then show that βT is a universal point transitive flow for T (in a sense which we shall make precise) and that a canonical action of βT is naturally defined on any flow (T,X). The algebraic structure of βT will then be related (via this action) to different dynamical properties of the flow X.

We recall that βT is a compact Hausdorff space which contains T as a dense subset and has the following universal property. Every map ϕ of T into a compact Hausdorff space X, can be extended to a continuous map $\phi: \beta T \to X$.

For a fixed $t \in T$ consider the map $s \to ts$. Thinking of it as a map from T into βT we can extend it to a continuous map $p \to tp$ of βT onto itself. Since T is discrete this gives us an action of T on βT. Clearly $(T,\beta T)$ is a point transitive flow, for example e has a dense orbit.

Next we consider, for a fixed $p \in \beta T$, the map $t \to tp$. This again can be extended to all of βT and we thus get the continous map $q \to qp$ of βT into itself. We note that the orbit closure of a point p in βT is given by $(\beta T)p = \{qp | q \in \beta T\}$. We now have a binary operation $\beta T \times \beta T \to \beta T$ which restricted to $T \times T$ is the group product. This gives a semigroup structure on βT. A subset E of βT is called a left ideal if $(\beta T)E \subset E$.

2.1 LEMMA: *A subset M of βT is a left minimal ideal of the semigroup βT iff it is a minimal subset of the flow $(T,\beta T)$. In particular a left minimal ideal is closed.*

Proof: This is an easy corollary of the fact that $\bar{\mathcal{O}}(p) = (\beta T)p$ for every $p \in M$. //

From now on we shall deal only with left minimal ideals of βT. Thus we shall refer to them as *minimal ideals.*

2.2. LEMMA: *Let E be a compact Hausdorff topological space provided with a semigroup structure such that the maps $y \to yx$ are continuous, for all $x \in E$. Then there exists an idempotent in E (i.e. an element x such that $x^2 = x$).*

Proof: Let $\mathcal{S}$ be the collection of all closed nonempty subsets S of E with the property $S^2 \subset S$. $\mathcal{S}$ is not empty because $E \in \mathcal{S}$. Now by Zorn's lemma there exists a minimal element in $\mathcal{S}$ (under inclusion), say S. If $x \in S$ then Sx is closed non-empty and $(Sx)(Sx) \subset S^3x \subset Sx$. Hence $Sx \in \mathcal{S}$ and since $Sx \subset S$ it follows that $Sx = S$.

Let $W = \{y \in S \mid yx = x\}$, then W is closed non-empty and clearly $W^2 \subset W \subset S$. Hence $W = S$, and $x^2 = x$ is an idempotent.//

2.3. PROPOSITION: *Let* M *be a minimal ideal of* βT, *and let* J *be the set of idempotents in* M. *Then*

(1) $J \neq \emptyset$

(2) *If* $v \in J$ *and* $p \in M$ *then* $pv = p$.

(3) *For each* $v \in J$, $vM = \{p \in M \mid vp = p\}$ *is a subgroup of* M, *with identity element* v. *The map* $p \to wp$ *is a group isomorphism of* vM *onto* wM *for every idempotent* $w \in J$.

(4) $\{vM \mid v \in J\}$ *is a partition of* M.

Proof: (1) This follows from lemma 2.2.

(2) Mv is a minimal ideal which is contained in M. Hence $Mv = M$ and if $p \in M$ then there exists $q \in M$ such that $qv = p$. Now $pv = (qv)v = qv^2 = qv = p$.

(3) We have to show that each element of vM has an inverse in vM. Let $p \in vM$ then as in (2) $Mp = M$. Hence there exists $q \in M$ such that $qp = v$. Also $Mq = M$ and there exists $r \in M$ such that $rq = v$. Now

$$p = vp = rqp = rv = r \qquad \text{and}$$

$$vq = qpq = qrq = qv = q.$$

Thus $q \in vM$ and $qp = pq = v$. We denote $q = p^{-1}$.

To see that $p \to wp$ from vM into wM is an isomorphism onto we observe that

$$[w(p^{-1})](wp) = wp^{-1}p = wv = w.$$

Thus $w(p^{-1}) = (wp)^{-1}$, also $w(pq) = (wp)wq$ and $v(wp) = vp = p$.

(4) If $p \in M$ then $Mp = M$. Hence the set $A = \{q \in M \mid qp = p\}$ is closed and non-empty and $A^2 \subset A$. By lemma 2.2 there is an idempotent $w \in A$ and thus $p \in wM$, i.e. $M = \bigcup\{vM \mid v \in J\}$. If $p \in vM \cap wM$ then $w = pp^{-1} = v$, hence the union is disjoint. //

2.4. COROLLARY: If we choose an arbitrary idempotent $u \in J$ and denote $G = uM$, then every element p of M has unique representation $p = v\alpha$ for $v \in J$ and $\alpha \in G$. Moreover $p^{-1} = v\alpha^{-1}$.

2.5. PROPOSITION: Let K, L and M be minimal ideals of βT. Let v be an idempotent in M. Then there is a unique idempotent v' in L such that $vv' = v'$ and $v'v = v$. (We write $v \sim v'$ and say that v' is equivalent to v). If v'' in K is equivalent to v' then $v \sim v''$. The map $p \to pv'$ of M onto L is an isomorphism of flows.

Proof: Let $v \in M$ be an idempotent then $Lv = M$ and by lemma 2.2 we conclude that the set $\{q \in L \mid qv = v\}$ contains an idempotent v'. Similarly $Mv' = L$ and we conclude the existence of an idempotent say $v_1 \in M$ such that $v_1v' = v'$. Now

$$v = v'v = v_1v'v = v_1v = v_1,$$

and thus $v \sim v'$. This also shows that v' is unique.

If $v' \sim v$ and $v'' \sim v'$ then

$$vv'' = v(v'v'') = (vv')v'' = v'v'' = v''$$

and similarly $v''v = v$. Thus $v'' \sim v$.

Finally it is clear that $p \to pv'$ is a flow homomorphism of M onto L. Since $p' \to p'v$ is also a homomorphism (of L into M) and $(pv')v = pv = p$ these homomorphisms are isomorphisms. //

Unless $M = L$ $p \to pv'$ will not be a semigroup isomorphism of M onto L.

2.6. PROPOSITION: The flow $(T,\beta T)$ is a universal point transitive flow, i.e. for every point transitive flow (T,X) and a point $x_o \in X$ such that $\mathcal{O}(x_o) = X$, there is a unique homomorphism $\beta T \xrightarrow{\phi} X$ such that $\phi(e) = x_o$.

Proof: We define $\phi(t) = tx_o$. This map can be extended to a continuous map ϕ of βT onto X. If $\{t_i\}$ is a net in T such that in βT $\lim t_i = p$ and $t \in T$ then

$$\begin{aligned}\phi(tp) &= \lim \phi(tt_i) = \lim tt_ix_o = t \lim t_ix_o \\ &= t\,\phi(p).\end{aligned}$$

Thus ϕ is a homomorphism and clearly $\phi(e) = x_o$. The uniqueness is also clear. //

Motivated by the last proposition we define a pointed flow (T,X,x_o) to be a flow with a distinguished point $x_o \in X$. A homomorphism of pointed flows $(X,x_o) \xrightarrow{\phi} (Y,y_o)$ is a homomorphism which satisfies $\phi(x_o) = y_o$. Thus the pointed flow $(T,\beta T,e)$ is universal among the pointed flows whose distinguished point has a dense orbit.

For a pointed flow (X,x_o) and $p \in \beta T$ we shall write $px_o = \phi(p)$ where $(\beta T,e) \xrightarrow{\phi} (X,x_o)$. If $x \in X$ is another point and $\mathcal{O}(x) = Y \subset X$ then there is a homomorphism $(\beta T,e) \xrightarrow{\psi} (Y,x)$ and we write $px = \psi(p)$. This gives us an "action" of βT on every flow. Clearly $p \to px$ is continuous but in general $x \to px$ will not be continuous. Notice that $(pq)x = p(qx)$, for all $p,q \in \beta T$ and $x \in X$.

This action of βT can be thought of as a representation of

βT in X^X, the compact space of all mappings of X into itself. The image of βT there, is a semigroup which is the closure of T (represented itself as a subset of X^X) in X^X. This semigroup is the <u>enveloping</u> (or the <u>Ellis</u>) <u>semigroup</u> of the flow X. We denote this semigroup by $E(X)$.

I.3. THE ACTION OF βT ON X.

<u>3.1. PROPOSITION</u>: *Let X be a flow and $x \in X$.*

(1) $\bar{\mathcal{O}}(x) = (\beta T)x$

(2) $\bar{\mathcal{O}}(x)$ *is minimal iff for every minimal ideal* $M \subset \beta T$, $x \in Mx$; *iff in every minimal ideal there is an idempotent* v *such that* $vx = x$.

<u>Proof</u>: (1) is clear; to prove (2) we observe that Mx is always a minimal set. Now if $\bar{\mathcal{O}}(x) = (\beta T)x$ is a minimal then $(\beta T)x = Mx$ and $x = ex \in (\beta T)x = Mx$. Finally if $x = vx$ for v an idempotent in M then $x = vx \in Mx$, and conversely if $x \in Mx$ then the set $\{p \in M | px = x\}$ contains an idempotent by lemma 2.2. //

<u>3.2. PROPOSITION:</u>

(1) *Let X be a flow. The following conditions are equivalent.*

(i) $(x,y) \in P(X)$

(ii) *There exists an element* $p \in \beta T$ *such that* $px = py$.

(iii) *There is a minimal ideal* M *such that* $px = py$ *for every* $p \in M$.

(2) *If X is minimal then for* $x \in X$

$$P(x) = \{y \in X \mid (x,y) \in P\}$$

$$= \{vx \mid v \text{ is an idempotent in some minimal ideal of } \beta T\}.$$

(3) *If* X *is minimal and* v *is an idempotent in some minimal ideal of* βT, *then every pair of points in* $vX = \{x \mid vx = x\}$ *is distal.*

Proof: (1) Clearly (i) and (ii) are equivalent and (iii) implies (ii). Suppose $px = py$ for some $p \in \beta T$. Let M be a minimal ideal in βT, then $N = Mp$ is also a minimal ideal and clearly $qx = qy$ for every $q \in N$.

(2) If v is an idempotent in βT and $x \in X$, then $(x,vx) \in P$, since $vx = v(vx)$. Now if X is minimal and $(x,y) \in P$ then for some minimal ideal M, $px = py$ for every $p \in M$. There is an idempotent v in M such that $vy = y$ (proposition 3.1) and $vx = vy = y$.

(3) If $x,y \in vX$ then $v(x,y) = (x,y)$ and hence $\bar{\mathcal{O}}(x,y)$ in $X \times X$ is a minimal set (proposition 3.1). If $(x,y) \in P$ this minimal set must be the diagonal. //

3.3. THEOREM: (1) $P(X)$ *is transitive iff* $E(X)$, *the enveloping semigroup, contains a unique minimal ideal.*

(2) X *is distal iff* $E(X)$ *is a group.*

(3) *If* X *is distal then* X *is the disjoint union of minimal sets.*

(4) X *is equicontinuous iff* $E(X)$ *is a topological group consisting of homeomorphisms.*

(5) X *is minimal and equicontinuous iff* $E(X)$ *is a topological group and the flow* (T,X) *is isomorphic to the homogeneous flow* $(T,E(X)/H)$, *where* H *is a closed subgroup of* $E(X)$ *and* $t(pH) = (tp)H$, $(t \in T,\ p \in E(X))$.

Proof: (1) Suppose P is transitive. Let u,u' be two equivalent idempotents, and let $x \in X$. Since (x,ux) and $(x,u'x)$ are in P so is $(ux,u'x)$. Now ux, $u'x$ are in uX hence by proposition 3.2 (3) they are distal and therefore $ux = u'x$. This is true for any $x \in X$, thus u and u' are equal as elements of $E(X)$. This implies that $E(X)$ contains a unique minimal ideal.

Conversely suppose that $E(X)$ contains a unique minimal ideal, then if (x,y) and (y,z) are in P, $px = py = pz$ for every p in this minimal ideal, hence $(x,z) \in P$.

(2) Let X be distal and let v be an idempotent in a minimal ideal M of βT. Then x and vx are both proximal and distal, hence $vx = x$, for every $x \in X$. Thus $v = e$ on X, and this implies that $E(X)$ is a homomorphic image of vM which is a group by proposition 2.3. Therefore $E(X)$ is a group. Conversely, if v is an idempotent in M, then the image of v in $E(X)$ is also an idempotent and hence the identity. Then $vX = X$ and proposition 3.2 (3) applies.

(3) This is clear from the fact that $E(X)$ is a group and that $\bar{\mathcal{O}}(x) = (\beta T)x = E(X)x$.

(4) The "only if" part is an easy consequence of the Arzelà-Ascoli theorem. For the other half you need Ellis's theorem on joint continuity (see [8,Theorems 1 and 3]).

(5) If X is minimal and equicontinuous then the action of the compact topological group $E(X)$ on X, is transitive and X is a homogeneous space. Clearly the converse is true. //

I.4. POINTED MINIMAL FLOWS AND THE ELLIS GROUP.

Let us fix from now on a minimal ideal M in βT. We denote by J its set of idempotents and we choose a distinguished idempotent $u \in J$. Denote by G the group uM. Following Ellis we denote the elements of G by greek letters α, β etc.

Given a minimal flow X, we choose a point $x_o \in uX = \{ux \mid x \in X\} = \{x \mid ux = x\}$. Under the canonical map $(\beta T, e) \to (X, x_o)$, M is mapped onto X and u onto x_o. Thus (M,u) is a <u>universal minimal pointed flow</u> in the sense, that for every minimal flow X there is a point $x_o \in X$ such that (X,x_o) is a factor of (M,u). Unless we say otherwise the base point x_o of a minimal flow (X,x_o) will be choosed so that $ux_o = x_o$.

Let (X,x_o) be a pointed minimal flow we define the Ellis group of (X,x_o) to be

$$\mathcal{G}(X,x_o) = \{\alpha \in G \mid \alpha x_o = x_o\}.$$

Clearly $\mathcal{G}(X,x_o)$ is a subgroup of G.

We say that a homomorphism $X \xrightarrow{\phi} Y$ is <u>proximal</u> (<u>distal</u>) if whenever $x_1, x_2 \in \phi^{-1}(y)$ then x_1 and x_2 are proximal (distal).

<u>4.1. PROPOSITION</u>: <u>Let</u> $(X,x_o) \xrightarrow{\phi} (Y,y_o)$ <u>be a homomorphism of pointed minimal flows</u>.

(1) $\mathcal{G}(X,x_o) \subset \mathcal{G}(Y,y_o)$.

(2) $\mathcal{G}(X,x_o) = \mathcal{G}(Y,y_o)$ <u>iff</u> ϕ <u>is proximal and if</u> ϕ <u>is proximal then</u> $\phi^{-1}(y) \subset Jx$ <u>for any</u> $x \in \phi^{-1}(y)$.

(3) ϕ <u>is distal iff for every</u> $y \in Y$ <u>and</u> $p \in M$ <u>such that</u> $py_o = y$

$$\phi^{-1}(y) = p\,\mathcal{G}(Y,y_o)x_o.$$

<u>Proof</u>: (1) Let $\alpha \in \mathcal{G}(X,x_o)$, then

$$\alpha y_o = \alpha\phi(x_o) = \phi(x_o) = y_o.$$

(This follows from the commutativity of the diagram

$$\begin{array}{ccc} (M,u) & & \\ \downarrow & \searrow^{\psi} & \\ (X,x_o) & \xrightarrow[\phi]{} & (Y,y_o) \end{array}$$

which in turn follows from the uniqueness of the map ψ.)

(2) Suppose the groups are equal. We show that $x_1, x_2 \in \phi^{-1}(y)$ implies x_1 and x_2 are proximal. So let $x_1 = px_o$, $x_2 = qx_o$ for p and q in M. Denote $\alpha = up^{-1}q$ then

$$\begin{aligned} \alpha y_o &= up^{-1}q\phi(x_o) = up^{-1}\phi(qx_o) = up^{-1}\phi(px_o) \\ &= up^{-1}p\phi(x_o) = uy_o = y_o . \end{aligned}$$

Thus $\alpha \in \mathcal{G}(Y,y_o)$ and hence $\alpha \in \mathcal{G}(X,x_o)$, i.e. $up^{-1}qx_o = x_o$. Hence

$$ux_1 = upx_o = up(up^{-1}qx_o) = ux_2,$$

and x_1 and x_2 are proximal.

Conversely suppose that ϕ is proximal and let $\alpha \in \mathcal{G}(Y,y_o)$. Then $\phi(\alpha x_o) = \alpha y_o = y_o$ implies αx_o and x_o are proximal. On the other hand by proposition 3.2 (3) αx_o and x_o are distal hence $\alpha x_o = x_o$ and $\alpha \in \mathcal{G}(X,x_o)$. Let $x, x_1 \in \phi^{-1}(y)$ then x and x_1 are proximal and by proposition 3.2 (3) there exists an idempotent v' in some minimal ideal L of βT such that $x_1 = v'x$. Now let $v \in J$ be such that $v \sim v'$, then $y = \phi(v'x) = \phi(vv'x) = v\phi(v'x) = vy$ and hence $\phi(vx) = v\phi(x) = vy = y$. It follows that vx and $v'x$ are proximal. But $vv'x = v'x$ and thus $vx, v'x \in vX$ and by 3.2 (3) vx and $v'x$ are also distal. Therefore $x_1 = v'x = vx \in Jx$ and the proof is completed.

(3) Suppose $\phi^{-1}(py_o) = p\mathcal{G}(Y,y_o)x_o$, and let $v \in J$ be such that

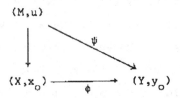

$vp = p$. Then $p\mathcal{G}(Y,y_0)x_0 \subset vX$ and ϕ is distal by 3.2 (3).

Let ϕ be distal. If $y = py_0$ for some $p \in M$ then for $\alpha \in \mathcal{G}(Y,y_0)$ we have

$$\phi(p\alpha x_0) = \phi(px_0) = p\phi(x_0) = py_0 = y.$$

Thus $p\mathcal{G}(Y,y_0)x_0 \subset \phi^{-1}(y)$. On the other hand if $\phi(x) = y$ then $x = qx_0$ for some $q \in M$ and since $\phi(qx_0) = y = \phi(px_0)$ we conclude as in (2), that $\alpha = up^{-1}q \in \mathcal{G}(Y,y_0)$. If $v \in J$ is such that $vq = q$ then $q = vp\alpha$. Now

$$y = \phi(x) = \phi(qx_0) \quad \text{and} \quad y = py_0 = p\alpha y_0 = \phi(p\alpha x_0).$$

Thus $qx_0 = v(p\alpha x_0)$ and $p\alpha x_0$ are both distal and proximal. Hence they are equal and $x = qx_0 = p\alpha x_0 \in p\mathcal{G}(Y,y_0)x_0$. //

<u>4.2. PROPOSITION</u>: <u>Let $(X,x_0) \xrightarrow{\phi} (Y,y_0)$ and $(Z,z_0) \xrightarrow{\psi} (Y,y_0)$ be two distal homomorphisms of minimal flows. There exists a homomorphism $(Z,z_0) \xrightarrow{\theta} (X,x_0)$ iff $\mathcal{G}(X,x_0) \supset \mathcal{G}(Z,z_0)$.</u>

<u>Proof</u>: If θ exists then $\mathcal{G}(X,x_0) \supset \mathcal{G}(Z,z_0)$ by 4.1 (1).

Suppose $\mathcal{G}(X,x_0) \supset \mathcal{G}(Z,z_0)$; for $p \in M$ define $\theta(pz_0) = px_0$. If $z = pz_0 = qz_0$ for $p,q \in M$ then $up^{-1}q \in \mathcal{G}(Z,z_0)$ and by our assumption $up^{-1}qx_0 = x_0$ and $pp^{-1}qx_0 = vqx_0 = px_0$ where $v = pp^{-1}$ is an idempotent in M. Thus qx_0 and px_0 are proximal. Now $pz_0 = qz_0$ implies $py_0 = qy_0$ and hence $\phi(px_0) = \phi(qx_0)$. This implies that px_0 and qx_0 are also distal i.e. $px_0 = qx_0$ and θ is well defined. Clearly θ is a continuous homomorphism and the proof is complete. //

Let $\{(X_\lambda,x_\lambda)\}_{\lambda\in\Lambda}$ be a family of pointed minimal flows, let z_0 be the point in the product space $\Pi_{\lambda\in\Lambda} X_\lambda$ which is defined by $(z_0)_\lambda = x_\lambda$ and let $Z = \bar{\mathcal{O}}(z_0)$. We denote (Z,z_0) by $\vee_{\lambda\in\Lambda}(X_\lambda,x_\lambda)$

4.3. LEMMA: Then (Z,z_0) is a minimal flow and $\mathcal{J}(Z,z_0) = \cap_{\lambda\in\Lambda}\mathcal{J}(X_\lambda,x_\lambda)$

Proof: Since $ux_\lambda = x_\lambda$ for each λ, $uz_0 = z_0$ and by 3.1 (2), Z is minimal. Now

$$\mathcal{J}(Z,z_0) = \{\alpha \in G \mid \alpha z_0 = z_0\} = \{\alpha \in G \mid \alpha x_\lambda = x_\lambda \quad \forall \lambda\}$$

$$= \bigcap_{\lambda\in\Lambda}\{\alpha \in G \mid \alpha x_\lambda = x_\lambda\} = \bigcap_{\lambda\in\Lambda}\mathcal{J}(X_\lambda,x_\lambda). \;//$$

CHAPTER II

PROXIMAL FLOWS

One would expect proximal flows to have very little in common with distal flows. Indeed we show next, that minimal proximal flows are disjoint from minimal distal flows, that they are weakly mixing, and that many "nice" groups do not admit at all such non-trivial flows. A theorem (due to Furstenberg) about the isomorphism of the universal minimal proximal flows of a group and its subsgroups of finite order is proved and several examples are introduced. More about the family of minimal flows which are disjoint from all minimal proximal flows and some examples can be found in [21].

II.1 PROXIMAL AND DISTAL FLOWS

1.1 LEMMA: *Let* $X \xrightarrow{\phi} Y$ *be a proximal homomorphism. If* Y *is minimal, then* X *contains a unique minimal set.*

Proof: Let X_1 and X_2 be minimal sets in X. Then $\phi(X_1) = Y = \phi(X_2)$, therefore there exists $x_i \in X_i$ $(i = 1,2)$, such that $\phi(x_1) = \phi(x_2)$. This implies that x_1 and x_2 are proximal and hence $X_1 = X_2$.//

1.2. LEMMA: *Let* $X \xrightarrow{\phi} Y$ *be a distal homomorphism. If* Y *is minimal, then* X *is a disjoint union of minimal sets.*

Proof: Let $x \in X$. Since Y is minimal there exists an idempotent v in M such that $v\phi(x) = \phi(x)$. But $\phi(vx) = v\phi(x) = \phi(x)$ and it follows that x and vx are distal. Clearly they are also proximal and hence $x = vx \in Mx$. Since Mx is a minimal subset of X, our lemma is proved.//

1.3. COROLLARY: *Every minimal proximal flow is disjoint from every minimal distal flow.*

Proof: Let X be minimal distal and let Y be minimal proximal. Then $X \times Y$ is a proximal extension of X (via the projection on the first coordinate) and a distal extension of Y (via the projection on the second coordinate). By lemmas 1.1 and 1.2 $X \times Y$ contains a unique minimal set and also it is the union of minimal sets, hence it is minimal.//

II.2 PROXIMAL AND WEAKLY MIXING FLOWS.

A flow X is *topologically ergodic* if its only closed invariant subset with a non-empty interior is X itself. X is *weakly mixing* if $X \times X$ is ergodic. The flows X and Y are *weakly disjoint* if $X \times Y$ is ergodic. (Thus X is weakly mixing iff it is weakly disjoint from itself. See [37]).

We have already defined the proximal relation P as the subset of $X \times X$ of points whose orbit closure intersects the diagonal. Let $P^{(n)}$ be the subset of X^n of all points whose orbit closure intersects the diagonal of X^n. Thus $P = P^{(2)}$.

2.1. PROPOSITION: *Let X be a minimal flow. If $P^{(n)}$ is dense in X^n for every $n \geq 2$, then X is weakly disjoint from all minimal flows.*

Proof: Let Y be a minimal flow and choose $W \subset X \times Y$ so that $\overline{W} = W$, W is T-invariant and $\mathrm{int}(W) \neq \emptyset$. Let U and V be open subsets of X and Y respectively, such that $U \times V \subset W$. The minimality of Y implies the existence of a finite subset $F = \{t_1, \cdots, t_n\}$ of T such that $Y = \bigcup_{i=1}^{n} t_i V$.

Let $Q = t_1U \times t_2U \times \cdots \times t_nU$, then Q is an open subset of

X^n and by our assumption there exists $(x_1,x_2,\cdots,x_n) \in Q \cap P^{(n)}$. Since X is minimal, for every $x \in X$, a net $\{s_j\}$ in T can be found such that

$$\lim s_j(x_1,x_2,\cdots,x_n) = (x,x,\cdots,x).$$

We choose now $y \in Y$; for every j the point $s_j^{-1}y$ belongs to at least one of the sets t_iV, $i = 1,\cdots,n$. Thus there exists an i_o, $1 \leq i_o \leq n$, for which $s_j^{-1}y \in t_{i_o}V$ for a co-final set of j's. Passing to this subnet we have $\lim s_j(x_1,x_2,\cdots,x_n) = (x,x,\cdots,x)$ and $s_j^{-1}y \in t_{i_o}V$ for all j.

Now $(x_{i_o}, s_j^{-1}y) \in t_{i_o}(U \times V) \subset t_{i_o}W = W$ hence

$$(s_jx_{i_o},y) = s_j(x_{i_o},s_j^{-1}y) \in s_jW = W$$

and taking the limit we have $(x,y) \in W$. Since x and y were arbitrary we conclude that $W = X \times Y$ i.e., X and Y are disjoint. //

2.2. COROLLARY: *Every minimal proximal flow is weakly disjoint from every minimal flow. In particular every minimal proximal flow is weakly mixing.*

Proof: Clearly $P^{(n)}$ is dense in X^n for X proximal. Thus our statement follows from proposition 2.1. //

II.3 STRONGLY AMENABLE GROUPS.

We say that a topological group is *strongly amenable* if every minimal and proximal flow of T is necessarily trivial. The reason for that name will become clear in Chapter III.

3.1. LEMMA: *Let T be a topological group, $S \subset T$ a closed subgroup such that T/S is compact. Then, if (T,X) is proximal so* *is* (S,X).

Proof. Let x_1, x_2 be points of X, and $\{t_i\}$ a net in T such that $\lim t_i(x_1,x_2) = (x,x)$ for some $x \in X$. In the compact space T/S we can assume that $\lim t_iS = tS$ for some $t \in T$. Then, there exists a net $\{s_i\}$ in S such that that $\lim t_is_i = t$. Again we can assume that the limits $z_1 = \lim s_i^{-1}x_1$ and $z_2 = \lim s_i^{-1}x_2$ exists and thus

$$(x,x) = \lim t_i(x_1,x_2) = \lim t_is_i(s_i^{-1}x_1, s_i^{-1}x_2) =$$

$$= t(z_1,z_2).$$

Hence $z_1 = z_2$ and x_1 and x_2 are S-proximal.//

We say that the topological group T is a *compact extension* of its closed subgroup S if S is normal in T and T/S is a compact homogeneous space.

3.2. *LEMMA: Let* T *be a topological group and* (T,X) *a minimal proximal flow.*

(1) *If* T *is a compact extension of* S *then* (S,X) *is minimal and proximal.*

(2) *If* S *is a subgroup of* T *of finite index in* T, *then* (S,X) *is minimal and proximal.*

Proof. (1) We know already that (S,X) is proximal, hence it contains a unique minimal set (lemma 1.1), say Y. If $t \in T$ and $s \in S$ then $st = ts'$ for some $s' \in S$. Thus $stY = ts'Y = tY$. If follows that tY is also S-minimal and hence $tY = Y$. We conclude that Y is T invariant and therefore $X = Y$, so that X is also S-minimal.

(2) Again (S,X) is proximal and contains a unique S-minimal set Y. Now there exists a finite subset $F = \{t_1,\cdots,t_n\}$ of T such

that $FS = T$. Hence $FY = X$ and it follows that the interior of Y is not empty. Since (S,X) is proximal for every $x \in X$ there is a $s \in S$ such that $sx \in Y$ but then $x \in s^{-1}Y = Y$ and we conclude that $Y = X$.//

Let (T,X) be a flow, a homomorphism of X into itself is called an endomorphism. If an endomorphism is one-to-one and onto it is called an automorphism.

3.3. LEMMA: Let (T,X) be a minimal flow. Let $X \xrightarrow{\phi} X$ be an endomorphism of X. Then for every $x \in X$, x and $\phi(x)$ are distal.

Proof: Let $x \in X$ and suppose $x \neq \phi(x)$ and x and $\phi(x)$ are proximal. Let $\{t_i\}$ be a net in T such that $\lim t_i x = \lim t_i\phi(x) = z$. Then $z = \lim t_i\phi(x) = \lim \phi(t_i x) = \phi(z)$. Since X is minimal z can be any point of X and this is a contradiction to our assumption that $x \neq \phi(x)$.//

3.4. THEOREM: Let T be a compact extension of S and suppose S is nilpotent. Then T is strongly amenable.

Proof: Let (T,X) be a minimal proximal flow. By lemma 3.2 so is (S,X). Now S is nilpotent it has a non-trivial center say S_1. Each element of S_1 acts as an automorphism of (S,X) and lemma 3.3 together with the proximatily of (S,X) implies that each of the elements of S_1 acts as the identity map on X. Thus the nilpotent group S/S_1 acts proximally and minimally on X. Our theorem follows now by induction on a central series of S.//

II.4 THE UNIVERSAL PROXIMAL FLOW.

For every topological group T we construct a universal minimal proximal flow; i.e., a minimal proximal flow which has every minimal proximal flow as a factor.

4.1. LEMMA: *The only endomorphism a minimal proximal flow admits is the identity automorphism.*

Proof: This is a consequence of lemma 3.3.//

4.2. PROPOSITION: *For every topological group* T, *there exists a universal minimal proximal flow. This flow is unique up to an isomorphism.*

We denote this flow by $(T,\pi(T))$.

Proof: Let X_i be the set of all non-isomorphic minimal proximal flows. Let $X' = \Pi X_i$ be the product flow. X' is a proximal flow and contains a unique minimal subset X. Clearly (T,X) is a universal minimal proximal flow. If Y is also universal minimal proximal flow for T then there are homomorphisms $X \xrightarrow{\phi} Y \xrightarrow{\psi} X$. Let $\chi = \psi\circ\phi$ then χ is an endomorphism of X hence, by lemma 4.1, χ = identity. Thus ϕ is an isomorphism.//

We note that *a topological group* T *is strongly amenable iff* $\pi(T)$ *is trivial*.

Let T be a topological group, denote by $A(T)$ the group of topological automorphisms of T and let $H(T)$ be the group of homeomorphisms of $\pi(T)$. We observe that lemma 4.1 implies that the action of the center of T on a minimal proximal flow is trivial.

4.3. PROPOSITION (Furstenberg): *There is a homomorphism of* $A(T)$ *into* $H(T)$ *which send the inner-automorphism* $\sigma_t\colon s \to tst^{-1}$ *onto the homeomorphism* $x \to tx$ *of* $\pi(T)$. *The action of* T *on* $\pi(T)$ *is effective* (i.e $tx = x$ *for all* x *implies* $t = e$) *iff this* homomorphism is one-to-one.

Proof: Let α be an automorphism of T. We define a new flow on the space $\pi(T)$ as follows: for $t \in T$ and $x \in \pi(T)$ let $t \cdot x = \alpha(t)x$. Clearly this flow is minimal and proximal; hence by the universality of $(T,\pi(T))$ there exists a map $\hat{\alpha}: \pi(T) \to \pi(T)$ such that

$$\hat{\alpha}(tx) = t \cdot \hat{\alpha}(x) = \alpha(t)\hat{\alpha}(x).$$

In the same way a map $\widehat{\alpha^{-1}}: \pi(T) \to \pi(T)$ is defined such that

$$\widehat{\alpha^{-1}}(tx) = \alpha^{-1}(t)\widehat{\alpha^{-1}}(x).$$

Now the composition map $\alpha \circ \widehat{\alpha^{-1}}$ satisfies

$$\alpha\circ\widehat{\alpha^{-1}}(tx) = \hat{\alpha}(\alpha^{-1}(t)\widehat{\alpha^{-1}}(x)) = \alpha\circ\alpha^{-1}(t)\hat{\alpha}\circ\widehat{\alpha^{-1}}(x)$$

$$= t\hat{\alpha}\circ\widehat{\alpha^{-1}}(x).$$

Thus $\hat{\alpha}\circ\widehat{\alpha^{-1}}$ is an endomorphism of $(T,\pi(T))$ and thus (lemma 4.1) it is the identity map. Therefor $\hat{\alpha}$ is a homeomorphism and $(\hat{\alpha})^{-1} = \widehat{\alpha^{-1}}$. If $\alpha,\beta \in A$ then one shows similarly that $\alpha\circ\beta\circ\widehat{(\alpha\beta)^{-1}}$ is an endomorphism of $\pi(T)$ and thus it is the identity map. Thus $\hat{\alpha}\circ\hat{\beta} = \widehat{\alpha\circ\beta}$ and $\alpha \to \hat{\alpha}$ is a homomorphism.

We now have to show that $\widehat{\sigma_t}(x) = tx$. For every $s \in T$

$$(\widehat{\sigma_t}\circ t^{-1})(sx) = \widehat{\sigma_t}(t^{-1}sx) = t(t^{-1}s)t^{-1}\widehat{\sigma_t}(x)$$

$$= st^{-1}\widehat{\sigma_t}(x) = s(t^{-1}t^{-1}t)\widehat{\sigma_t}(x)$$

$$= s\widehat{\sigma_t}(t^{-1}x) = s(\widehat{\sigma_t}\circ t^{-1})(x).$$

Thus $\widehat{\sigma_t}\circ t^{-1}$ is an endomorphism of $\pi(T)$ hence equal to the identity and $\widehat{\sigma_t} = t$ on $\pi(T)$. The last statement is clear. //

It follows from proposition 4.2 that the flow $(T,\pi(T))$ can be extended to a flow $(A(T),\pi(T))$. However, usually in the latter flow, we can not take $A(T)$ to be a topological group in its compact open topology.

4.4. THEOREM (Furstenberg). *Let* T *be a discrete group,* S *a subgroup of finite index in* T. *Then the action of* S *on* $\pi(S)$ *can be extended to an action of* T *on* $\pi(S)$ *such that* $(T,\pi(T))$ *and* $(T,\pi(T))$ *are isomorphic*. (*In particular* $\pi(T)$ *and* $\pi(S)$ *are homeomorphic*.)

Proof: First we choose a normal subgroup of T, say N, which is contained in S and has a finite index in T, (See [27, page 26] for example.) We can define an action of T on $\pi(N)$ by mapping T into $A(N)$ ($t \to \sigma_t$ where $\sigma_t(n) = tnt^{-1}$) and then extending $(N,\pi(N))$ to $(A(N),\pi(N))$. We thus have a flow $(T,\pi(N))$ which is clearly minimal and proximal. By the universality of $(T,\pi(T))$ there is a homomorphism $(T,\pi(T)) \xrightarrow{\phi} (T,\pi(N))$.

Next we consider the flow $(N,\pi(T))$ which is obtained by restricting the action of T on $\pi(T)$ to the action of its subgroup N. By lemma 3.2(1) this flow is minimal and proximal; therefor there is a homomorphism $(N,\pi(N)) \xrightarrow{\psi} (N,\pi(T))$. Now $\phi\circ\psi$ is an endomorphism of $(N,\pi(N))$ and hence it is the identity. It follows that ϕ is a T-isomorphism.

By lemma 3.2(2) $(S,\pi(T))$ is minimal and proximal hence there is a homomorphism $(S,\pi(S)) \xrightarrow{\theta} (S,\pi(T))$; similarly $(N,\pi(S))$ is minimal and proximal and there exists a homomorphism $(N,\pi(N)) \xrightarrow{\lambda} (N,\pi(S))$. Thus the composition

$$(N,\pi(N)) \xrightarrow{\lambda} (N,\pi(S)) \xrightarrow{\theta} (N,\pi(T)) \xrightarrow{\phi} (N,\pi(N))$$

is an endomorphism of $(N,\pi(N))$ and hence it is the identity. We conclude that $(N,\pi(S))$ is isomorphic to $(N,\pi(T))$ and using this isomorphism an action of T on $\pi(S)$ can be defined so that $(T,\pi(S))$ and $(T,\pi(T))$ are isomorphic.//

II.5 EXAMPLES OF PROXIMAL FLOWS.

5.1. Let $X = \{0,1\}^Z$, and let t be the shift on X, $(t\xi)_i = \xi_{i-1}$. We define the following homeomorphisms of X.

$$s_0(\xi) = \begin{cases} \xi & \text{if } \xi_0 = 0 \\ \xi' & \text{if } \xi_0 = 1 \text{ where } \xi_i' = \xi_i \text{ iff } i \neq 1 \end{cases}$$

and,

$$s_1(\xi) = \begin{cases} \xi & \text{if } \xi_0 = 1 \\ \xi' & \text{if } \xi_0 = 0. \end{cases}$$

Let T be the group of homeomorphisms of X generated by t, s_0 and s_1. We claim that (T,X) is minimal and proximal.

To see that X is minimal we observe that the homeomorphism $s_1 s_0$ changes the first coordinate and nothing else. Thus starting with an element ξ of X and using iterates of t and then $s_1 s_0$, we can change any coordinate of ξ. Clearly this implies that the orbit of ξ is dense.

We show now that X is proximal. Let $\varepsilon,\eta \in X$, if $\varepsilon_0 \neq \eta_0$ and $\varepsilon_1 \neq \eta_1$ then by applying either s_0 or s_1 we get $\varepsilon_1' = \eta_1'$. Thus we can use an element r of T to get a block of length one common to $r\varepsilon$ and $r\eta$. Let us assume that ε and η coincide on some block B of length n, say $\varepsilon_j = \eta_j$ $j = 2,3,\cdots,n+1$. If $\varepsilon_1 = \eta_1$ we are done, if not let i be the largest integer smaller than 1, for which $\varepsilon_i \neq \eta_i$. If $i = 0$ we get equality at the first coordinate by applying either s_0 or s_1. If $i < 0$, by using iterates of the shift, then either s_0 or s_1 and then shifting back we get inequality at the $i+1$ coordinate. We repeat this to get inequality at the zero coordinate and then the first coordinate can be made identical enlarging the length of B by one.

5.2. Let $Y = \{e^{2\pi i\theta} \mid 0 \leq \theta < 1\}$, let $Y \xrightarrow{t} Y$ be given by $t(e^{2\pi i\theta}) = e^{2\pi i\theta^2}$, and $Y \xrightarrow{s} Y$ by $s(e^{2\pi i\theta}) = e^{2\pi i(\theta+\beta)}$. Let T be the group of homeomorphism of Y generated by t and s. If β is irrational then (T,Y) is minimal. It is clear that under iterates of t every point of Y is proximal to 1, thus (T,Y) is proximal.

5.3. Let F be the free group on two generators a and b. Let $X' = \{a,b,a^{-1},b^{-1}\}^{\mathbb{N}}$, the set of all unilateral sequences on the symbols a, b, a^{-1}, b^{-1}. Let $A_n = \{\xi \in X' \mid \xi(n) = \xi(n+1)^{-1}\}$ with the understanding that $(a^{-1})^{-1} = a$ and $(b^{-1})^{-1} = b$. Let $A = \bigcup_{n=1}^{\infty} A_n$ and let $X = X' \setminus A$; X is closed, hence compact, subset of X'.

Define an action of F on X as follows. If w is a reduced word in F and ξ is an element of X we define w to be this element of X which is obtained by concatenation of w and ξ and then performing the necessary cancellations. This action is well defined and continuous because there can be at most n cancellations, where n is the length of w.

Let $\xi,\eta \in X$, and let η starts with the word w of length n. Suppose w ends with the letter ε and ξ starts with the letter δ; $\varepsilon,\delta \in \{a,b,a^{-1},b^{-1}\}$. We can always choose a letter ν so that $\varepsilon\nu\delta$ is reduced. Let $w' = w\nu$ then $w'\xi$ and η have identical first n letters, thus (F,X) is minimal.

If ξ and η in X starts with the letters δ and ε respectively then there is a letter ν so that both $\nu\delta$ and $\nu\varepsilon$ are reduced. We let $w = \nu\nu\cdots\nu$ (n times) then $w\xi$ and $w\eta$ have identical first n letters. Thus (F,X) is also proximal.

Similar examples can be constructed with F replaced by any free group or a free product of groups.

5.4. We can consider the flow (T,Y) of 5.2 as a flow for F, thinking of T as a homomorphic image of F via the map $a \to s$, $b \to t$. We claim that (F,Y) and (F,X), where the latter is the flow of example 5.3., are disjoint. Indeed the orbit closure of the point $(a_\infty,y) \in X \times Y$ under the subgroup $A = \{a^n \mid n \in Z\}$ of F, is $\{a_\infty\} \times Y$, where a_∞ is the element $aaa\cdots$ of X and y is an arbitrary point in Y. Let Z be the unique minimal set of the proximal flow $X \times Y$ then Z contains some point of the form (a_∞,y) and hence $\{a_\infty\} \times Y \subset Z$. This implies $Z = X \times Y$ and $(F,X\times Y)$ is a minimal proximal flow.

5.5. (Furstenberg) Let Φ be a field which contains the rational numbers (i.e. chr $\Phi = 0$). Let V be a vector space over Φ, and consider V as a discrete abelian group. We let $\hat{V}$ be its compact abelian dual group. Since Φ^* (the multiplicative group of Φ) acts as a group of automorphisms of V it acts also as a group of automorphisms of $\hat{V}$. This makes $\hat{V}$ a vector space over Φ.

Let U be a dense subgroup of $\hat{V}$ and denote

$$X = \left\{\begin{pmatrix} x \\ 1 \end{pmatrix} \mid x \in \hat{V}\right\}$$

$$S = \left\{\begin{pmatrix} \alpha & x \\ 0 & 1 \end{pmatrix} \mid \alpha \in \Phi^*, \quad x \in U\right\}.$$

Then X is a compact subset of $\hat{V} \oplus \Phi$, which is invariant under the action of the solvable group of matrices, S. The flow (S,X) is clearly minimal.

If x is in $\hat{V}$ then the set

$$\text{cls } \{nx \mid n \in \Phi^* \text{ a positive integer}\}$$

is a compact subsemigroup of $\hat{V}$. Hence by lemma I.2.2, this semigroup contains an idempotent which must be 0. We can conclude that the orbit closure of $\begin{pmatrix} x \\ 1 \end{pmatrix}$ under the action of the matrices in S of the form $\begin{pmatrix} n & 0 \\ 0 & 1 \end{pmatrix}$, contains $\begin{pmatrix} 0 \\ 1 \end{pmatrix}$. But each of these matrices fixes $\begin{pmatrix} 0 \\ 1 \end{pmatrix}$, hence $\begin{pmatrix} x \\ 1 \end{pmatrix}$ and $\begin{pmatrix} 0 \\ 1 \end{pmatrix}$ are proximal and if follows that (S,X) is proximal.

In particular if we take $V = \Phi = \mathbb{R}$, then $\hat{V}$ is the Bohr compactification of $\mathbb{R}$ and we can take $U = \mathbb{R}$ to be a dense sub-group of $\widehat{\mathbb{R}_d}$. Now S is the group of matrices

$$S = \left\{ \begin{pmatrix} a & b \\ 0 & 1 \end{pmatrix} \middle| \; a \in \mathbb{R}^*, \quad b \in \mathbb{R} \right\}$$

which is the group of affine transformations of the real line. Thus <u>there exist solvable groups which are not strongly amenable</u>.

Notive however that the action of S on X is <u>not</u> continuous in the group variable when S is equipped with its usual topology, because $\mathbb{R}_d$ contains non-continuous characters of the reals with their usual topology.

5.6. Let $\mathbb{P}^1$ be the projective line, i.e. the set of lines through the origin of the plane. Let $\mathbb{G} = \mathbb{S}\mathbb{L}_2(\mathbb{R})$ be the group of real 2×2 matrices with determinant 1. $\mathbb{G}$ acts naturally on $\mathbb{P}^1$ (sending lines into lines), and this action is transitive. If we denote by x_o the line which contains the vector $\begin{pmatrix} 1 \\ 0 \end{pmatrix}$; i.e., the real axis, then

$$\mathbb{S} = \{g \in \mathbb{G} \mid gx_o = x_o\} = \left\{ \begin{pmatrix} a & b \\ 0 & a^{-1} \end{pmatrix} \mid a \neq 0 \right\}.$$

Since $\mathbb{G} = \mathbb{K} \cdot \mathbb{S}$ where $\mathbb{K}$ is the compact group of rotations we conclude that the flor $(\mathbb{G},\mathbb{P}^1)$ is isomorphic to the homogeneous flow $(\mathbb{G},\mathbb{G}/\mathbb{S})$ and that $\mathbb{K}$ is already transitive on $\mathbb{P}^1$.

To see that this flow is proximal we consdier the sequence of matrices $\{g_n\}_{n=0}^{\infty}$

$$g_n = \begin{pmatrix} 1 & n \\ 0 & 1 \end{pmatrix}.$$

For every line $x \in \mathbb{P}^1$, $\lim g_n x = x_o$.

CHAPTER III

STRONGLY PROXIMAL FLOWS

The existence of an invariant measure on a flow is a very desirable property of the flow. However when dealing with general acting groups an invariant measure does not necessarily exist. Looking for an invariant measure on a flow (T,X) is the same as looking for a fixed point in the flow $(T,M(X))$, where $M(X)$ is the space of probability measures on X.

We consider the family of closed invariant <u>convex</u> subsets of $M(X)$ and choose the minimal ones among them. If we are lucky, we find an invariant measure. In any case these minimal objects turn out to be strongly proximal flows. Section two is devoted to the proof of this theorem, which is due to Furstenberg.

There is a strong connection between the existence of strongly proximal flows and amenability, this is studied in section 3. In section 4 we give a proof due to W.A. Veech of the existence of 2^c invariant means for a countable discrete amenable group.

Some fixed point theorems are proved in section 5, and in section 6 it is shown that a minimal flow with an invariant measure is disjoint from every minimal strongly proximal flow. We consider some examples in section 7.

III.1 DEFINITIONS

Let X be a compact Hausdorff space, let $C(X)$ denote the Banach algebra of all real valued continuous functions on X with the supremum norm. Let $C^*(X)$ be the dual space of $C(X)$ and let $M(X)$ be the set of the regular Borel probability measures on X. We can think of $M(X)$ as a subset of $C^*(X)$, writing $\mu(f) = \int_X f\,d\mu$ for $\mu \in M(X)$ and $f \in C(X)$.

With the weak $*$ topology induced from $C^*(X)$, $M(X)$ is a compact Hausdorff convex space. We recall that a net $\mu_i \in M(X)$ converges in this topology to the measure $\mu \in M(X)$ iff $\int_X f\,d\mu_i \to \int_X f\,d\mu$ for every $f \in C(X)$.

If (T,X) is a flow we can associate with each element t of T an isometry, $f \to f^t$ of $C(X)$ onto itself, where $f^t(x) = f(tx)$. We now define an action of T on the space $M(X)$ as follows. For $\mu \in M(X)$ and $t \in T$ let $t\mu$ be the linear functional on $C(X)$ defined by:

$$\int_X f\,d(t\mu) = (t\mu)f = \mu(f^t) = \int_X f^t\,d\mu.$$

$(f \in C(X))$.

Clearly each t maps $M(X)$ onto itself in a continuous and one-to-one way. Thus $\mu \to t\mu$ is a homeomorphism. Moreover if $t_i \to t$ and $\mu_i \to \mu$ then $t_i\mu_i \to t\mu$. Hence $(T,M(X))$ is a flow. An important property of this flow is the fact that each $t \in T$ acts affinely on $M(X)$; i.e., $t(a\mu+(1-a)\nu) = at\mu + (1-a)t\nu$ for all $\mu,\nu \in M(X)$ and $0 \le a \le 1$.

The flow (T,X) is said to be <u>strongly proximal</u> if the flow $(T,M(X))$ is proximal. With every $x \in X$ we have the point mass $\delta_x \in M(X)$. The map $x \to \delta_x$ is an isomorphism of the flow X onto a subflow of $M(X)$. We shall identify X with this subflow. Using this isomorphism it is clear that strong proximality implies proximality.

<u>1.1 LEMMA</u>: X <u>is strongly proximal iff given</u> $\mu \in M(X)$ <u>there exists a net</u> $\{t_i\}$ <u>in</u> T <u>such that</u> $\lim t_i\mu = x$ <u>for some</u> $x \in X$.

<u>Proof</u>: ($\Rightarrow$) This implication is clear because X is a closed invariant subset of $M(X)$ and thus the unique minimal subset of the proximal flow $M(X)$ is contained in X.

(<=) Let $\mu, \nu \in M(X)$; denote $\theta = \frac{1}{2}(\mu + \nu)$. Then there exists a net $\{t_i\}$ in T such that $\lim t_i\theta = x$. Now we can assume that $\lim t_i\mu = \mu_1$ and $\lim t_i\nu = \nu_1$ exist and then $x = \frac{1}{2}(\mu_1 + \nu_1)$. This implies that $\mu_1 = \nu_1 = x$ because x is an extreme point of $M(X)$. //

It is easy to see that the product of strongly proximal flows is strongly proximal, and that a subflow of a strongly proximal flow is strongly proximal. Since every strongly proximal flow is also proximal, a minimal strongly proximal flow does not admit non-trivial endomorphisms. (Lemma II.4.1). Hence as in proposition II.4.2 one proves that there exists a unique, up to an isomorphism, universal minimal strongly proximal flow. We denote this flow by $\pi_S(T)$.

In the same way proposition II.4.3 and theorem II.4.4 can be restated for π_S instead of π.

Motivated by the example of the flow $(T,M(X))$ we now define the notion of an affine flow.

III.2 AFFINE FLOWS

Let E be a real locally convex topological vector space, let Q be a compact convex subset of E and let (T,Q) be a flow. If each $t \in T$ acts on Q as an affine transformation, then we say that (T,Q) is an affine flow. An affine flow is said to be irreducible if it contains no proper closed convex invariant set. When $Q \subset E$ is compact and convex, we use the following notations. $ex(Q)$ is the set of extreme points in Q and $\overline{ex}(Q)$ is the closure of this set. If X is a subset of Q we write $\overline{co}(X)$ for the closed convex hull of X. For the proofs of the following propositions we refer to [38].

2.1. PROPOSITION: Let $Q \subset E$ be a compact and convex set. There exists a unique map (called the barycenter map) $\beta: M(Q) \to Q$, satisfying the following conditions.

(a) β is onto.

(b) β is affine.

(c) β is weak * continuous.

(d) If f is a continuous real valued affine function on Q then

$$f(\beta(\mu)) = \int_Q f \, d\mu \qquad (\mu \in M(Q)).$$

We note that if (T,Q) is an affine flow then in addition to the properties (a)-(d), β is a homomorphism of flows. For

$$f(t\beta(\mu)) = f^t(\beta(\mu)) = \int_Q f^t \, d\mu =$$

$$= \int_Q f \, d(t\mu) = f(\beta(t\mu)),$$

and the affine functions separate points in Q.

2.2. PROPOSITION: (1) Let X be a closed subset of Q such that $\overline{co}(X) = Q$. Then $ex(Q) \subset X$.

(2) If $x \in ex(Q)$ then δ_x, the point mass at x, is the unique measure μ in $M(Q)$ such that $\beta(\mu) = x$.

2.3. THEOREM: (Furstenberg) Let (T,Q) be an irreducible affine flow. Then (T,Q) is a strongly proximal flow. If $X = \overline{ex}(Q)$, then X is the unique minimal set of Q. Thus (T,X) is a minimal strongly proximal flow.

Proof: Since $ex(Q)$ is an invariant subset of Q, X is a closed invariant subset. Let $x \in Q$ then by the irreducibility of Q, $\overline{co}(\overline{\mathcal{O}}(x)) = Q$ and hence by proposition 2.2., $X \subset \overline{\mathcal{O}}(x)$. It follows that X is the unique minimal set in (T,Q).

Now let $\mu \in M(Q)$ and let $\beta: M(Q) \to Q$ be the barycenter map. Then $\beta(\overline{co}(\mathcal{O}(\mu))) = Q$ because it is a closed convex invariant subset of Q. Note that $\beta(\overline{co}(\mathcal{O}(\mu))) = \overline{co}(\beta(\mathcal{O}(\mu)))$ and hence $X \subset \beta(\mathcal{O}(\mu))$. If $x \in ex(Q) \subset X$ then by proposition 2.2.(2) $\beta^{-1}(x) = \{\delta_x\}$ and thus $\delta_x \in \overline{\mathcal{O}}(\mu)$. By lemma 1.1. (T,Q) is strongly proximal.//

2.4. PROPOSITION: $(T,M(\pi_S(T)))$ is a universal irreducible affine flow; i.e., for every irreducible affine flow (T,Q) there exists an affine homomorphism of $M(\pi_S(T))$ onto Q. Moreover this homomorphism is unique.

Proof: Clearly $M(\pi_S(T))$ is irreducible. Let (T,Q) be an irreducible affine flow and let X be its unique minimal set. By Theorem 2.3. X is a minimal strongly proximal flow, hence there exists a unique homomorphism $\pi_S(T) \to X$. Let $M(\pi_S(T)) \xrightarrow{\Phi} M(X) \subset M(Q)$ be the induced map and let $M(Q) \xrightarrow{\beta} Q$ be the barycenter map. Then $\beta \circ \Phi$ is the unique affine homomorphism of $M(\pi_S(T))$ onto Q.//

III.3 AMENABLE GROUPS

Given a topological group, we denote by L the Banach algebra of all real valued bounded left uniformly continuous functions on T, with the supremum norm. (A function f is left uniformly continuous (L.U.C.) on T if for every $\varepsilon > 0$ there exists a neighborhood V of $e \in T$ such that $\|f - f^t\| < \varepsilon$ for every $t \in V$.)

A mean m on L is a continuous, positive, linear functional such that $m(1) = 1$. A mean is called invariant if $m(f^t) = m(f)$ for every $f \in L$ and $t \in T$. T is called amenable if there exists an invariant mean on L. T has the fixed point property if whenever (T,Q) is an affine flow then T has a fixed point in Q. (See [23]).

3.1. THEOREM: Let T be a topological group, the following are equivalent.

(1) T *is amenable*.

(2) *Every flow* (T,X) *has an invariant measure*.

(3) T *has the fixed point property*.

(4) *Every minimal strongly proximal flow for* T *is trivial* (*or equivalently* $\pi_S(T)$ *is trivial*.)

Proof: (1) <=> (2) (A skech of a proof). Let $|L|$ be the compact Hausdorff space of maximal ideals (multiplicative functionals) of the Banach algebra L. Then L is isometricaly isomorphic to $C(|L|)$. The isometries $f \to f^t$ of L induce a group of homeomorphisms $x \to tx$ of $|L|$. This action of T on $|L|$ is jointly continuous and thus $(T,|L|)$ is a flow. Moreover the functionals x_t: $f \to f(t)$ $(t \in T)$ form a dense subset of $|L|$ and clearly for $t,s \in T$, $sx_t = x_{st}$. Thus x_e has a dense orbit in $|L|$ and $(T,|L|)$ is point transitive.

If (X,x_o) is a pointed flow, we define a map $L_{x_o} : C(X) \to L$ by

$$(L_{x_o} f)(t) = f(tx_o) \qquad (f \in C(X), t \in T).$$

If $\mathcal{O}(x_o) = X$, then L_{x_o} is an isometry of $C(X)$ into L. The maximal ideal space $|A|$ of the closed T-invariant subalgebra A, which is the image of $C(X)$ under L_{x_o} can be made into a flow $(T,|A|)$ and this flow is isomorphic to (T,X). The restriction of a functional in $|L|$ to A is a homomorphism of $(T,|L|,x_e)$ onto $(T,|A|,x_e)$, which in turn is isomorphic to (T,X,x_o).

Thus $(T,|L|,x_e)$ is a universal point transitive flow for the topological group T. The existence of an invariant mean on L is clearly equivalent to the existence of an invariant, positive, continuous, linear functional μ on $C(|L|)$ with $\mu(1) = 1$, i.e., an invariant probability measure on $|L|$.

Now, if every flow (T,X) has an invariant measure so does $|L|$ and hence T has an invariant mean. On the other hand if (T,X) is an arbitrary flow, we choose a point $x \in X$ and let $Y = \bar{\mathcal{O}}(x)$. Then (T,Y) is a point transitive flow, hence it is a homomorphic image of $|L|$ and if $|L|$ has an invariant measure so does Y and hence also X.

(2) => (3) Let (T,Q) be an affine flow and let Q_o be an irreducible affine subflow. (This always exists by Zorn's lemma.) By (2) (T,Q_o) has an invariant measure, but by theorem 2.3. (T,Q_o) is strongly proximal. Clearly this implies that Q_o contains just one point which is a fixed point for T in Q.

(3) => (4) Let (T,X) be a minimal strongly proximal flow. By our assumption $M(X)$ contains a fixed point and by the strong proximality this point must be a point of X. Since X is minimal it must be trivial.

(4) => (2) Let (T,X) be a flow we consider the affine flow $(T,M(X))$ and an irreducible subflow (T,Q). By theorem 2.3. Q is strongly proximal hence by our assumption it is trivial, thus the unique point of Q is an invariant measure for (T,X).//

Since $\pi_S(T)$ is a minimal proximal flow we now see that <u>strong amenability</u> (i.e., $\pi(T)$ is trivial) <u>implies amenability</u> ($\pi_S(T)$ is trivial).

3.2. THEOREM: <u>The following kinds of groups are amenable.</u>

(a) <u>Abelian groups</u>.

(b) <u>Compact groups</u>.

(c) <u>Solvable groups</u>.

(d) <u>Compact extensions of solvable groups</u>.

<u>Proof</u>: The groups of types (a) and (b) were already shown to be strongly amenable, hence they are amenable.

Let T be a solvable group, with a derived series $\{e\} = T_o \subset T_1 \subset \cdots \subset T_n = T$. We show that T has the fixed point property. Let (T,Q) be an affine flow and consider the affine flow (T_1,Q). Since T_1 is abelian, the set Q_1 of fixed points of T_1 is a convex closed non-empty subset of Q. Since T_1 is normal in T_2, Q_1 is also T_2-invariant. Now the effective action of T_2 on Q_1 is abelian and hence the subset Q_2 of Q_1 which consists of T_2-fixed points is closed convex non-empty and T_3-invariant. The proof is completed by induction.

Finally if S is a solvable, normal subgroup of T and T/S is compact, then in every affine flow (T,Q) the set Q_o of S-fixed points is closed, convex, non-empty and T-invariant. Since the effective action of T on Q_o is compact, T has a fixed point in Q_o. //

3.3. LEMMA: *Let T be a topological group, S a closed co-compact amenable subgroup. If $(T,T/S)$ has an invariant measure then T is amenable.*

Proof: Since S is amenable there is an S-fixed point x_o in the affine flow $(S,M(\pi_S(T)))$. The map $tS \xrightarrow{\phi} tx_o$ is a homomorphism of $(T,T/S)$ onto a subflow of $(T,M(\pi_S(T)))$. Let m be a T-invariant probability measure on T/S; then $\phi(m)$ is a T-invariant point in $M(\pi_S(T))$. This implies that $\pi_S(T)$ is trivial; i.e., T is amenable. //

III.4 COUNTABLY INFINITE AMENABLE GROUP HAS 2^c DIFFERENT INVARIANT MEANS

Let T be a discrete group; the algebra L coincides with the Banach algebra $B(T)$ of all bounded functions on T and the latter can be identified with the algebra $C(\beta T)$ of continuous functions on the Stone-Chêch compactification of T. It follows that there is a one-to-one correspondence between invariant means on L and invariant probability measures on βT.

We show that if T is infinite countable and discrete then βT contains 2^c minimal ideals. If moreover T is amenable then each of these minimal ideals carries an invariant measure and these correspond to 2^c different means on L.

4.1. THEOREM: (Veech) *Let T be a countably infinite amenable group, then βT has 2^c minimal ideals.*

Proof: Let $T_1 \subset T_2 \subset \cdots$ be a sequence of finite subsets of T such that $e \in T_1$ and $\bigcup_{n=1}^{\infty} T_n = T$. Choose elements $t_1, t_2, \cdots$ in T such that $T_n t_n \cap T_m t_m \neq \emptyset$ for $n \neq m$.

If E is an infinite subset of $\mathbb{N}$, define sets $A(E)$, $B(E) \subset \beta T \setminus T$ by

$$A(E) = \overline{\bigcup_{n \in E} T_n t_n} \setminus \bigcup_{n \in E} T_n t_n$$

$$B(E) = \overline{\bigcup_{n \in E} \{t_n\}} \setminus \bigcup_{n \in E} \{t_n\}.$$

If $p \in B(E)$, then there exists a net $\{n_i\}$ in E such that $p = \lim t_{n_i}$. If $t \in T$, then $tt_{n_i} \in T_{n_i} t_{n_i}$ for all large i and hence $tp = \lim tt_{n_i} \in A(E)$. It follows that $A(E) \supset \overline{Tp}$; thus $A(E)$ contains a closed invariant set.

Now fix a point $a \in \beta\mathbb{N} - \mathbb{N}$, and let $S_a = \{E \in \mathbb{N} \mid E$ is infinite and $a \in \overline{E}\}$. Define

$$A(a) = \cap \{A(E) \mid E \in S_a\}$$

$$B(a) = \cap \{B(E) \mid E \in S_a\}.$$

If $p \in B(a)$, then $tp \in A(a)$ as above and thus $A(a)$ contains a T-invariant set. If a and b are in $\beta\mathbb{N} \setminus \mathbb{N}$ and $a \neq b$ then there are sets $E \in S_a$ and $E' \in S_b$ such that $E \cap E' = \emptyset$ (This is true because the sets of the form $\overline{D}$ where $D \subset T$, form a basis for the topology of $\beta\mathbb{N}$. [Chapter 9, 8]) It follows that $A(E) \cap A(E') = \emptyset$ and hence also $A(a) \cap A(b) = \emptyset$. Since $A(a)$ contains a closed invariant set, it contains a minimal set. Thus the cardinality of minimal sets in βT is at least as great as the cardinality of $\beta\mathbb{N}\setminus\mathbb{N}$ which is 2^c. On the other hand $\beta T\setminus T$ has at most 2^c closed sets altogether.//

4.2. COROLLARY: Let T be a countably infinite amenable group. Then T has 2^c different invariant means.

III.5 FIXED POINT THEOREMS

Theorem 3.2. is the Markov-Kakutani fixed point theorem [7]. We proved it using theorem 2.3. Using 2.3. again we shall now prove the following two fixed point theorems.

5.1. THEOREM: (Hahn) [25]: Let (T,Q) be an affine flow. If the action of T on Q is distal then T has a fixed point in Q.

Proof: Choose an irreducible subset $Q_o \subset Q$; then (T,Q_o) is both distal and strongly proximal (theorem 2.3.) hence it is trivial and Q consists of just one point which is a fixed point for T.//

5.2. THEOREM: (Ryll Nardzewski) Let E be a separable Banach space, Q a weakly compact convex subset of E. Let (T,Q) be an affine flow such that the action of T is distal in the norm topology, then T has a fixed point in Q.

Proof: We can assume that (T,Q) is an irreducible affine flow. Let B be a closed ball of radius $\varepsilon > 0$ centered at the origin of E. Denote $X = \overline{ex}(Q)$ (closure in the weak topology). There exists a set $\{x_i\}_{i=1}^{\infty}$ of points in X such that $\{x_i + B\}_{i=1}^{\infty}$ is a denumerable cover of X. B is norm closed and convex, hence it is also weakly closed. Thus by Baire's category theorem we can find a weak neighborhood W of some x_i, such that $W \cap X \subset (x_i + B) \cap X$.

Since Q is irreducible, X is the unique minimal set of the flow Q and by theorem 2.3 it is strongly proximal. If $x,y \in X$ we can find a $t \in T$ such that tx and ty are in $X \cap W \subset x_i + B$. Since ε was arbitrary, this contradicts the distality in the norm of T, unless X and hence also Q are trivial. //

III.6 DISJOINTNESS

6.1. THEOREM: *Let (T,X) be a minimal flow with an invariant measure. Then X is disjoint from every minimal strongly proximal flow.*

Proof: We show that $X \times \pi_S(T)$ is minimal. Let W be a closed invariant subset of $X \times \pi_S(T)$, and let π_1 and π_2 be the projections Pick $\nu \in M(W)$ such that $\pi_1(\nu) = \mu$ is an invariant measure on X, and denote $\pi_2(\nu) = \theta$.

There exists a net $\{t_i\}$ in T such that $\lim t_i\theta = \delta_z$ for any $z \in \pi_S(T)$. We can assume that $\lim t_i\nu = \lambda$ exists. Then $\pi_1(\lambda) = \pi_1(\lim t_i\nu) = \lim t_i\pi_1(\nu) = \mu$ and $\pi_2(\lambda) = \pi_2(\lim t_i\nu) = \lim t_i\pi_2(\nu) = \delta_z$. It follows that $\lambda = \mu \times \delta_z$ and since $\lambda \in M(W)$, we can conclude that the support of λ, which is clearly $X \times \{z\}$, is contained in W. Since z is arbitrary the proof if completed. //

For a characterization of the family of minimal flows which are disjoint from $\pi_S(T)$, see [21].

III.7 PROXIMAL AND STRONGLY PROXIMAL

In this section we study the examples 5.1-5.6 of Chapter II to determine whether each of them is strongly proximal or not.

5.1. Take the measure which gives mass $1/2$ to each of the points of the two point space $\{0,1\}$, and form the corresponding product measure m on $X = \{0,1\}^{\mathbf{Z}}$. Clearly the homeomorphisms t, s_0 and s_1 of X preserve m. Thus the flow (T,X) has an invariant measure and therefore is an example of a minimal proximal flow which is not strongly proximal. Moreover by theorem 6.1. X is disjoint from $\pi_S(T)$.

5.2. If $z = e^{i\theta}$ is an arbitrary point of Y then $\lim_{n\to\infty} t^n z = 1$. Hence by the Lebesgue dominated convergence theorem

$$\int_Y f \, d(t^n\mu) = \int_Y f^{t_n} \, d\mu \to f(1) = \delta_1(f),$$

for every $f \in C(Y)$ and $\mu \in M(Y)$. Thus $\lim t^n\mu = \delta_1$ for every $\mu \in M(Y)$ and (T,Y) is strongly proximal.

5.3. Let a_∞ and a_∞^{-1} denote the elements $aaa\cdots$ and $a^{-1}a^{-1}a^{-1}\cdots$ respectively. Let $x \in X$, $x \neq a_\infty^{-1}$, then $\lim_{n\to\infty} a^n x = a$. Thus for a measure $\mu \in M(X)$ for which a_∞^{-1} is not an atom, we have again by the dominated convergence theorem $\lim a^n = \delta_{a_\infty}$.

If $\mu \in M(X)$ we can write $\mu = \alpha\delta_{a_\infty^{-1}} + (1-\alpha)\nu$ where $0 \le \alpha \le 1$ and $\nu \in M(X)$ is such that $\nu(\{a_\infty^{-1}\}) = 0$. Then $\lim a^n\mu = \alpha\delta_{a_\infty^{-1}} + (1-\alpha)\delta_{a_\infty} = \eta$ and for example $\lim b^n\eta = \delta_{b_\infty}$ where $b_\infty = bbb\cdots$. Thus (F,X) is strongly proximal.

5.4. In this example $X \times Y$ is also strongly proximal as a product of strongly proximal flows.

5.5. The acting group S in this example is solvable and in particular it is amenable. By theorem 3.1. $\pi_S(S)$ is trivial and hence (S,X) can not be strongly proximal. We conclude the amenability does not imply strong amenability.

5.6. Again the Lebesgue dominated convergence theorem can be used to show that $(T,\mathbb{P}^1)$ is strongly proximal.

CHAPTER IV
THE FURSTENBERG BOUNDARY OF A LIE GROUP

In [14] Furstenberg introduced the notion of a <u>boundary</u> of a Lie group, i.e., a strongly proximal homogeneous flow of the group. In chapter VI we shall show how the symmetric space D associated with a semisimple connected Lie group $\mathbb{G}$ with a finite center, can be represented as a subset of $M(X)$, where X is a maximal boundary for $\mathbb{G}$, is such a way that $X \subset \bar{D} \setminus D$. (Hence the name boundary).

In this chapter we show that for a connected Lie group, every strongly proximal flow is a boundary. Thus $\pi_S(\mathbb{G})$, the universal strongly proximal flow of $\mathbb{G}$, is isomorphic to a homogeneous flow $\mathbb{G}/\mathbb{H}$. We study the group $\mathbb{H}$ and in the case in which $\mathbb{G}$ is connected semisimple with a finite center we use a result of C.C. Moore, (which we state without proof) to identify $\mathbb{H}$ as the normalizer of $\mathbb{S}$ where $\mathbb{G} = \mathbb{K} \cdot \mathbb{S}$ is an Iwasawa decomposition of $\mathbb{G}$.

In section one we collect all the general results from Lie group theory that we shall need in these notes. In section two we bring the characterization, due to Furstenberg, of amenable connected Lie groups as the compact extensions of solvable groups. In section four we study boundaries of $SL(n,\mathbb{R})$. In section five some applications to disjointness are shown, and in section six the existence of proximal minimal flows for Lie groups is discussed. The main source for this chapter is [14].

IV.1 BASIC FACTS FROM LIE GROUP THEORY

1.1. <u>The adjoint group</u>

If V is a finite dimensional real vector space we denote by $\mathbb{GL}(V)$ $(g\ell(V))$ the Lie group (Lie algebra) of all automorphisms (endomorphisms) of V. $g\ell(V)$ is the Lie algebra of $\mathbb{GL}(V)$.

Let $\mathbb{G}$ be a connected Lie group, g its Lie algebra. For $\sigma \in g$ we denote by $ad(g)$ the endomorphism $\tau \to [\sigma,\tau]$ of g into itself. ad is a homomorphism of g into $g\ell(g)$. Let $Int(g)$ be the connected subgroup of $\mathbb{GL}(g)$ which corresponds to $ad(g)$. $Int(g)$ is called the adjoint group of g. For $g \in \mathbb{G}$ let $I(g)$ be the automorphism of $\mathbb{G}$ defined by $h \to ghg^{-1}$. To this automorphism corresponds an automorphism of g which is denoted by $Ad(g)$.

Ad <u>is an analytic homomorphism of</u> $\mathbb{G}$ <u>onto</u> $Int(g)$ <u>with kernel</u> Z (<u>where</u> Z <u>is the center of</u> $\mathbb{G}$). <u>The map</u> $gZ \to Ad(g)$ <u>is an analytic isomorphism of</u> $\mathbb{G}/Z$ <u>onto</u> $Int(g)$.

1.2. Semisimple Lie groups

If h_1 and h_2 are subalgebras of g, $[h_1,h_2]$ is the subalgebra of g generated by $\{[\alpha,\beta] \mid \alpha \in h_1, \beta \in h_2\}$. The <u>derived series</u> of g is the sequence ideals of g, $D^k g$ where $D^0 g = g$ and $D^k g = [D^{k-1}g, D^{k-1}g]$. g is called <u>solvable</u> if $D^k g = \{0\}$ for some k. $\mathbb{G}$ <u>is solvable iff</u> g <u>is solvable</u>. The <u>descending central series</u> of g is defined by $C^0 g = g$, $C^k g = [g, C^{k-1}g]$; g is nilpotent if for some k $C^k g = \{0\}$. $\mathbb{G}$ <u>is nilpotent iff</u> g <u>is nilpotent</u>.

The Lie algebra g is <u>semisimple</u> if it has no proper solvable ideals; g is <u>simple</u> if it is semisimple and has no ideals except $\{0\}$ and g. $\mathbb{G}$ is <u>semisimple</u> (<u>simple</u>) if g semisimple (simple). The <u>radical</u> of $\mathbb{G}$ is its maximal connected normal solvable subgroup, and $\mathbb{G}$ <u>is semisimple iff its radical is</u> $\{1\}$. Thus, if $\mathbb{S}$ is the radical of $\mathbb{G}$ then $\mathbb{G}/\mathbb{S}$ is semisimple.

For $\sigma,\tau \in g$ let $B(\sigma,\tau) = Tr(ad(\sigma)ad(\tau))$ where $Tr(A)$ is the trace of the endomorphism A. B is a symmetric bilinear form on g which is called the <u>Killing form</u> we have

$$B(\sigma,[\tau,\lambda]) = B(\tau,[\lambda,\sigma]) = B(\lambda,[\sigma,\tau]).$$

g <u>is semisimple iff</u> B <u>is nondegenerate</u>.

If $f\colon \mathbb{G} \to \mathbb{GL}(V)$ is a representation we say that f is <u>irreducible</u> if no proper subspace of V is invariant under $f(\mathbb{G})$. f is <u>completely reducible</u> if V is a direct sum of irreducible subspaces under f. *If* $\mathbb{G}$ *is semisimple then every representation of* $\mathbb{G}$ *is completely reducible*.

1.3. Compactness

The algebra g is called *compact* if $\mathrm{Int}(g)$ is a compact group. *If* $\mathbb{G}$ *is semisimple then* $\mathbb{G}$ *is a compact group iff* g *is compact*. For, if $\mathbb{G}$ is compact then so is its homomorphic image $\mathrm{Ad}(\mathbb{G}) = \mathrm{Int}(g)$. Conversely if g is compact then $\mathrm{Int}(g)$ is compact and by a theorem of Weyl [26, Chpt.II, Theorem 6.9] the universal covering group of $\mathrm{Int}(g)$ is compact. Since $\mathbb{G}$ is semisimple, its center Z is discrete and it follows that $\mathbb{G}$ and $\mathrm{Int}(g)$ have the same universal covering. Thus, *a semisimple connected Lie group* $\mathbb{G}$ *is compact iff* $\mathrm{Ad}(\mathbb{G})$ *is a compact group*.

1.4. Cartan decomposition.

If g is semisimple then there exists a *Cartan involution* $\theta: g \to g$; i.e., θ is an automorphism of g such that θ^2 = identity and $-B(\sigma,\theta(\tau))$ $(\sigma,\tau \in g)$ is a positive definite bilinear form on g. Any two Cartan involutions are conjugate under some $\mathrm{Ad}(g)$. We denote

$$k = \{\alpha \in g \mid \theta(\alpha) = \alpha\}$$

$$p = \{\alpha \in g \mid \theta(\alpha) = -\alpha\},$$

then $g = k \oplus p$; this decomposition is called a *Cartan decomposition* of g. If $\kappa \in k$ and $\pi \in p$ then $B(\kappa,\pi) = 0$. Since g is semisimple its ad representation is faithfull and one can represent g as a Lie algebra of matrices in such a way that all the elements of k are represented by skew symmetric matrices and those of p by symmetric matrices [36].

Let $\mathbb{K}$ be the connected subgroup of $\mathbb{G}$ which corresponds to k. Let $\psi: \mathbb{K} \times p$ be defined by $\psi(k,\alpha) = k \exp \alpha$, then ψ *is an* analytic isomorphism (of analytic manifolds) of $\mathbb{K} \times p$ onto $\mathbb{G}$. If $\kappa \in k$ and $\pi \in p$ then $\theta[\kappa,\pi] = [\theta\kappa,\theta\pi] = [\kappa,-\pi] = -[\kappa,\theta]$, thus $[\kappa,\pi] \in p$ and p is ad$k$ stable and hence also Ad($\mathbb{K}$) stable.

The subgroup $\mathbb{K}$ is compact iff $\mathbb{G}$ has a finite center. If $\mathbb{G}$ has a finite center, $\mathbb{K}$ is a maximal compact subgroup of $\mathbb{G}$ and for every maximal compact subgroup $\mathbb{K}$ of $\mathbb{G}$ there exists a Cartan decomposition $g = k \oplus p$ where k is the subalgebra of g which corresponds to $\mathbb{K}$.

1.5. Iwasawa decomposition

Let g be semisimple and $g = k \oplus p$ a Cartan decomposition. An abelian subalgebra a of g which is contained in p and is maximal with respect to these properties is called a Cartan subalgebra. For $\lambda \in a^*$, the dual space of a, let

$$g^\lambda = \{\alpha \in g \mid [\alpha,\alpha'] = \lambda(\alpha)\alpha' \text{ for all } \alpha' \in a\}.$$

Let $\Phi = \{\lambda \in a^* \mid g^\lambda \neq 0, \lambda \neq 0\}$ then g is the direct sum $g = g^0 + \sum_{\lambda \in \Phi} g^\lambda$. Choose a basis $\{\beta_1,\cdots,\beta_r\}$ for a and define a lexicographic order on a^* by setting $\lambda > 0$ if $\lambda \neq 0$ and if $i = \min\{k \mid \lambda(\beta_k) \neq 0\}$ then $\lambda(\beta_i) > 0$. Let $\Phi^+ = \{\lambda \in \Phi \mid \lambda > 0\}$ and $n = \sum_{\lambda \in \Phi^+} g^\lambda$, then n is nilpotent subalgebra of g.

Let $\mathbb{K}$, $\mathbb{A}$ and $\mathbb{N}$ be the connected subgroups of $\mathbb{G}$ which corresponds to k, a and n respectively then $\mathbb{A}$ and $\mathbb{N}$ are closed and the map $\mathbb{K} \times \mathbb{A} \times \mathbb{N} \to \mathbb{G}$ defined by $(k,a,n) \to kan$ is an analytic diffeomorphism onto $\mathbb{G}$. Denote $\mathbb{S} \doteq \mathbb{A} \cdot \mathbb{N}$ then $\mathbb{S}$ is a closed connected subgroup of $\mathbb{G}$ and $\mathbb{N}$ is normal in $\mathbb{S}$. We say that $\mathbb{G} = \mathbb{K}\mathbb{A}\mathbb{N}$ or $\mathbb{G} = \mathbb{K}\mathbb{S}$ is an Iwasawa decomposition of $\mathbb{G}$. [26, Chpt.VI, Theorem 5.1].

1.6. Unimodular groups

A locally compact group T is called unimodular if a right

invariant measure on $\mathfrak{G}$ is also left invariant. If $\mathfrak{G}$ is a Lie group then $\mathfrak{G}$ is unimodular iff $|\det \mathrm{Ad}(g)| = 1$ for all $g \in \mathfrak{G}$. [26, Chpt.10, Proposition 1.4].

If $\mathfrak{G}$ is a connected semisimple Lie group then the function $g \to \det \mathrm{Ad}(g)$ is a homomorphism of $\mathfrak{G}$ into $\mathbb{R}$. Since a homomorphic image of a semisimple group is semisimple, the image must be $\{1\}$. Thus _a semisimple Lie group is unimodular_.

_If_ T _is a locally compact topological group,_ S _a closed co-compact unimodular subgroup then the flow_ $(T,T/S)$ _has an invariant measure_. [43].

IV.2 CHARACTERIZATION OF AMENABLE LIE GROUPS

2.1. LEMMA: _Let_ $\mathfrak{G}$ _be a closed connected amenable Lie group of_ $m \times n$ _matrices, such that_ $|\det(g)| = 1$ _for all_ $g \in \mathfrak{G}$. _Suppose_ $\mathfrak{G}$ _acts irreducibly on_ $\mathbb{R}^m$, _then_ $\mathfrak{G}$ _is a compact group_.

Proof: The action of $\mathfrak{G}$ on $\mathbb{R}^m$ induces an action of $\mathfrak{G}$ on $\mathbb{P}^{m-1}$, the projective space of all lines through the origin of $\mathbb{R}^m$. For an element $v \in \mathbb{R}^m$ we denote by $\bar{v}$ its image in $\mathbb{P}^{m-1}$.

If $\mathfrak{G}$ is not compact, there exists a sequence of matrices $\{g_n\}$ in $\mathfrak{G}$ such that $\|g_n\| \to \infty$ (where $\|g\| = \max|g_{ij}|$). Let $h_i = g_i/\|g_i\|$, then $\|h_i\| = 1$ and we can assume that the sequence h_i converges to a matrix h. Now since $|\det(g_i)| = 1$ it follows that $\det(h) = 0$; i.e., h is a singular matrix. Let $N = \ker(h)$, $L = \mathrm{image}(h)$ and let N_1 and L_1 be the corresponding linear subvarieties of $\mathbb{P}^{m-1}$.

The space of all the r-dimensional subspaces of $\mathbb{R}^m$, where $r = \dim N$, is a compact space hence we can assume that in this space $\lim g_i(N) = Q$ exists. Thus Q is an r-dimensional subspace of $\mathbb{R}^m$ and if we denote by Q_1 the corresponding linear subvariety of $\mathbb{P}^{m-1}$, then $\lim g_i(N_1) = Q_1$.

Since $\mathfrak{G}$ is amenable there exists a $\mathfrak{G}$-invariant probability measure μ on $\mathbb{P}^{m-1}$. Write $\mu = \mu_1 + \mu_2$ where μ_1 is the restriction of μ to N_1. We can assume that $\nu_j = \lim g_i\mu_j$ for $j = 1,2$, exist and since μ is $\mathfrak{G}$-invariant we have $\mu = \nu_1 + \nu_2$. Clearly $g_i\mu_1$ is supported in g_iN_1 and hence the support of ν_1 is contained in Q_1.

Now for $v \in \mathbb{R}^m \setminus N$, $\lim \overline{g_i(v)} = \lim \overline{h_i(v)} = \overline{h(v)}$, hence for every $x \in \mathbb{P}^{m-1}\setminus N_1$, $\lim g_ix$ exists and it is an element of L_1. Let f be a positive continuous function on $\mathbb{P}^{m-1}$ which vanishes on L_1, then $\lim f^{g_i} = 0$ pointwise on $\mathbb{P}^{m-1}\setminus N_1$ and by the Lebesgue dominated convergence theorem

$$\int_{\mathbb{P}^{m-1}} f\, d\nu_2 = \lim \int_{\mathbb{P}^{m-1}} f\, d(g_i\mu_2) =$$

$$= \lim \int_{\mathbb{P}^{m-1}} f(g_ix)\, d\mu_2(x)$$

$$= \int_{\mathbb{P}^{m-1}\setminus N_1} (\lim f(g_ix))d\ {}_2(x) = 0.$$

Thus the support of ν_2 is contained in L_1; we conclude that μ is supported in $L_1 \cup Q_1$ and at least one of the subvarieties L_1 and Q_1 has a positive μ-measure. Let M be a subvariety of $\mathbb{P}^{m-1}$ of minimal dimension with a positive μ-measure. Then for $g,h \in \mathfrak{G}$ $gM \cap hM$ is either gM or a subvariety of a lower dimension and hence having measure zero. Since for each $g \in G$, $\mu(gM) = \mu(M) > 0$ this implies that there are only finitely many different translates of M. Since $\mathfrak{G}$ is connected $\mathfrak{G} = \mathfrak{G}' = \{g \in \mathfrak{G} \mid gM = M\}$ and M is $\mathfrak{G}$-invariant. Lifting M back to $\mathbb{R}^m$ we have a closed invariant proper subspace of $\mathbb{R}^m$, a contradiction to the irreducibility of $\mathfrak{G}$ on $\mathbb{R}^m$. //

<u>2.2 THEOREM</u> (Furstenberg). <u>If G is an amenable connected semi-simple Lie group then G is compact.</u>

Proof: By 1.3 it is enough to show that Ad(G) is a compact group. By 1.6. the matrices of Ad(G) are unimodular; by 1.2. Ad(G) is completely reducible on g, hence we can assume that it is irreducible. Since G is semi-simple, Ad(G) is a real semi-simple Lie group of matrices. This implies that Ad(G) is the identity component a real algebraic group and hence closed. Finally by our assumption Ad(G) is amenable. The theorem now follows from lemma 2.1.//

2.3. THEOREM: (Furstenberg): A connected Lie group is amenable iff it is a compact extension of a solvable group.

Proof: Let $\mathbb{G}$ be a connected amenable Lie group and let $\mathbb{S}$ be its radical. Then $\mathbb{G}/\mathbb{S}$ is semisimple connected and amenable. By theorem 2.2. $\mathbb{G}/\mathbb{S}$ is compact. The other direction of the theorem follows from III.3.2.//

IV.3. BOUNDARIES OF CONNECTED LIE GROUPS.

Let $\mathbb{G}$ be a connected semisimple Lie group with finite center and let $\mathbb{G} = \mathbb{K}\mathbb{A}\mathbb{N} = \mathbb{K}\mathbb{S}$ be an Iwasawa decomposition of $\mathbb{G}$. The reader can find the proof of the following proposition in [42, Proposition 1.2.3.4].

3.1. PROPOSITION: Let $\mathbb{P}$ be the normalizer of $\mathbb{N}$ in $\mathbb{G}$ and let $\mathbb{M}$ be the centralizer of $\mathbb{A}$ in $\mathbb{K}$, then $\mathbb{P} = \mathbb{M}\mathbb{A}\mathbb{N}$.

If we denote the normalizer of $\mathbb{S}$ in $\mathbb{G}$ by $\mathbb{H}$ then it follows from this proposition that $\mathbb{P} \subset \mathbb{H}$. Now consider the flow $(\mathbb{G},\pi_S(\mathbb{G}))$, since $\mathbb{S}$ is solvable it is amenable and there exsits an $\mathbb{S}$-invariant measure $\mu \in M$; since $\pi_S(\mathbb{G})$ is strongly proximal, $M(\pi_S(\mathbb{G}))$ is proximal and by II,3.1. the action of the co-compact subgroup $\mathbb{S}$ on $M(\pi_S(\mathbb{G}))$ is also proximal. Thus there exists a net $\{s_i\}$ in $\mathbb{S}$ such that $\lim s_i\mu$ is a point mass on $\pi_S(\mathbb{G})$. Since μ is $\mathbb{S}$ invariant we conclude that μ is a point mass, say $\mu = \delta_z$ and z is a fixed point for $\mathbb{S}$ in $\pi_S(\mathbb{G})$.

If $g \in \mathbb{H}$ and $s \in \mathbb{S}$, then there is some $s' \in \mathbb{S}$ such that $sg = gs'$ and $s(gz) = gs'z = gz$. Thus gz is also $\mathbb{S}$-fixed; but gz and z are $\mathbb{S}$ proximal and we conclude that $gz = z$ and that $\mathbb{H}$ has a fixed point in $\pi_S(\mathbb{G})$. Since $\mathbb{G}/\mathbb{H}$ is a compact space it follows that the minimal flow $(\mathbb{G},\pi_S(\mathbb{G}))$ is isomorphic to a homogeneous flow $(\mathbb{G},\mathbb{G}/\mathbb{L})$ where $\mathbb{L}$ is a closed subgroup containing $\mathbb{H}$.

Now C.C. Moore has shown that $(\mathbb{G},\mathbb{G}/\mathbb{P})$ is a strongly proximal flow (see [41, Prop.1.2.3.13] or [33 and 34]). By the universality of $\pi_S(\mathbb{G}) \cong \mathbb{G}/\mathbb{L}$ it follows that $\mathbb{P} \subset \mathbb{H} \subset \mathbb{L} \subset \mathbb{P}$. Thus we conclude that $\pi_S(\mathbb{G}) \cong \mathbb{G}/\mathbb{P}$ and that $\mathbb{P}$ is also the normalizer of $\mathbb{S}$ in $\mathbb{G}$ (i.e. $\mathbb{P} = \mathbb{H}$).

<u>3.2. THEOREM</u>: (Furstenberg-Moore). <u>Let $\mathbb{G}$ be a connected semisimple Lie group with a finite center. Then $(\mathbb{G},\pi_S(\mathbb{G}))$ is isomorphic to the boundary $(\mathbb{G},\mathbb{G}/\mathbb{P})$. Moreover for a closed subgroup $\mathbb{H}$ of $\mathbb{G}$, the homogeneous flow $\mathbb{G}/\mathbb{H}$ is a boundary iff $\mathbb{H}$ contains some conjugate of $\mathbb{P}$.</u>

<u>Proof</u>: We have to prove only the last assertion and this follows from the universality of $\mathbb{G}/\mathbb{P}$ by the existence of a homomorphism $(\mathbb{G},\mathbb{G}/\mathbb{P}) \xrightarrow{\phi} (\mathbb{G},\mathbb{G}/\mathbb{H})$. If we denote the coset $\{\mathbb{P}\}$ by x_0 and if $\phi(x_0) = g\mathbb{H}$ then $\mathbb{P}\phi(x_0) = \phi(x_0) = g\mathbb{H}$. Thus $\mathbb{P}g\mathbb{H} \subset g\mathbb{H}$ and $g^{-1}\mathbb{P}g \subset \mathbb{H}$. //

We remark that $\mathbb{P}$ is a maximal amenable subgroup of $\mathbb{G}$. Indeed since $\mathbb{S}$ is normal in $\mathbb{P}$, $\mathbb{P}$ is a compact extension of a solvable group and thus amenable. If $\mathbb{P} \subset \mathbb{H}$ and $\mathbb{H}$ is amenable, then the flow $(\mathbb{H},\mathbb{G}/\mathbb{P})$ has an invariant measure μ. Since $\mathbb{H}$ is co-compact in $\mathbb{G}$, $(\mathbb{H},\mathbb{G}/\mathbb{P})$ is strongly proximal and it follows that μ is a point mass on $\mathbb{G}/\mathbb{P}$. Thus $\mathbb{H}$ fixes a coset of $\mathbb{P}$, say $\mathbb{H}g\mathbb{P} = g\mathbb{P}$.

It follows that $g^{-1}\mathbb{H}g \subset \mathbb{P} \subset \mathbb{H}$. Now, since $\mathbb{K}\mathbb{P} = \mathbb{G}$ we have also $\mathbb{K}\mathbb{H} = \mathbb{G}$ and for some $k_o \in \mathbb{K}$ $k_o^{-1}\mathbb{H}k_o \subset \mathbb{P} \subset \mathbb{H}$. Let $\mathbb{K}_o = \{k \in \mathbb{K} \mid k^{-1}\mathbb{H}k \subset \mathbb{H}\}$ then $\mathbb{K}_o$ is a closed semigroup hence a subgroup of $\mathbb{K}$ (I.2.2.). Thus $k_o^{-1} \in \mathbb{K}_o$ i.e. $k_o\mathbb{H}k_o^{-1} = \mathbb{H}$ and hence $k_o^{-1}\mathbb{H}k_o = \mathbb{H}$. This implies $\mathbb{H} = \mathbb{P}$.

We say that a closed subgroup $\mathbb{H}$ of a connected Lie group $\mathbb{G}$ is a B-<u>subgroup</u> if $\mathbb{G}/\mathbb{H}$ is a boundary. If $\mathbb{H}$ is minimal with respect to this property it is called a <u>minimal</u> B-<u>subgroup</u>. Thus for a connected semisimple Lie group with a finite center the minimal B-subgroups of $\mathbb{G}$ are the conjugates of $\mathbb{P}$ where $\mathbb{G} = \mathbb{K}\mathbb{S}$ is an Iwasawa decomposition and $\mathbb{P}$ is the normalizer of $\mathbb{S}$ in $\mathbb{G}$. It follows therefore that all the $\mathbb{P}$'s which arise in this way are conjugate.

<u>3.3. THEOREM</u> (Furstenberg) <u>Let $\mathbb{G}$ be a connected Lie group with a radical</u> $\mathbb{S}$.

(1) <u>The universal minimal strongly proximal flow for</u> G <u>is a boundary</u>.

(2) <u>This boundary is trivial iff</u> $\mathbb{G}/\mathbb{S}$ <u>is compact</u>.

(3) <u>The minimal</u> B-<u>subgroups of</u> $\mathbb{G}$ <u>are conjugate</u>.

(4) <u>The minimal</u> B-<u>subgroups of</u> $\mathbb{G}$ <u>are maximal amenable subgroups</u>.

<u>Proof</u>: (1) The set of $\mathbb{S}$-fixed points in $M(\pi_S(\mathbb{G}))$ is closed convex $\mathbb{G}$-invariant non-empty subset. Since the affine flow is $\mathbb{G}$-irreducible it follows that this subset is all of $M(\pi_S(\mathbb{G}))$ and in particular, $\mathbb{S}$ acts trivially on $\pi_S(\mathbb{G})$. Thus we can consider $\mathbb{G}/\mathbb{S}$ (which is connected and semisimple) instead of $\mathbb{G}$. Now if Z is the center of $\mathbb{G}/\mathbb{S}$ then we show as above that the action of Z on $\pi_S(\mathbb{G})$ is trivial. Thus the effective action of $\mathbb{G}$ on $\pi_S(\mathbb{G})$ is really this of a connected semisimple Lie group with a trivial center and (1) follows.

(2) This follows form III.3.2. the proof of (1) and Theorem 3.2.

(3) and (4) follow easily from the semisimple with finite center case.//

IV.4 THE BOUNDARIES OF $\mathbf{SL}(n,\mathbb{R})$.

4.1. Let $\mathbb{G} = \mathbb{SL}(n,\mathbb{R})$ the group of $n \times n$ real matrices with determinant one. $\mathbb{G}$ is a connected semisimple Lie group with finite center. If we take $\mathbb{K} = \mathbf{SO}(n)$, the subgroup of $\mathbb{G}$ of orthogonal matrices, $\mathbb{A}$ the subgroup of diagonal matrices with positive entries on the diagonal and $\mathbb{N}$ the subgroup of uppertriangular matrices with 1's on the diagonal, then $\mathbb{G} = \mathbb{K}\mathbb{A}\mathbb{N}$ is an Iwasawa decomposition for $\mathbb{G}$. Clearly M, the centralizer of $\mathbb{A}$ in $\mathbb{K}$ is the subgroup of all diagonal matrices with ± 1 on the diagonal. Thus $\mathbb{P} = \mathbb{M}\mathbb{A}\mathbb{N}$, which is a minimal B-subgroup of $\mathbb{G}$, is the group of all upper triangular matrices in $\mathbb{G}$.

Let

$$F_n = \{(V_1,V_2,\cdots,V_{n-1}) \mid V_1 \subset V_2 \subset \cdots \subset V_{n-1} \text{ are subspaces of } \mathbb{R}^n \text{ and } \dim V_i = i\}$$

be the flag manifold with the usual compact topology. The action of $\mathbb{G}$ on $\mathbb{R}^n$ induces an action of $\mathbb{G}$ on F_n. If we denote $x_0 = (E_1,E_2,\cdots,E_{n-1})$ where $E_i = \{v \in \mathbb{R}^n \mid v_j = 0 \text{ for } j > i\}$ then the isotropy subgroup of $\mathbb{G}$ at x_0 is $\mathbb{P}$. Since the action of $\mathbb{G}$ on F_n is transitive it follows that $(\mathbb{G},F_n)$ is isomorphic to the homogeneous flow $(\mathbb{G},\mathbb{G}/\mathbb{P})$ which is the universal strongly proximal flow for $\mathbb{SL}(n,\mathbb{R})$.

$(\mathbb{G},F_n)$ has the Grassman variety $G_{i,n-1}$ of the i-dimensional subspaces of $\mathbb{R}^n$ with the natural action of $\mathbb{G}$ as a factor, thus all the Grassmanian flows $(\mathbb{G},G_{i,n-1})$ are boundaries. In particular $G_{1,n-1} = \mathbb{P}^{n-1}$, the projective n-1 space is a boundary for $\mathbb{G}$.

IV.5 DISJOINTNESS

The fact that $\pi_S(\mathbb{G})$ is a homogeneous space when $\mathbb{G}$ is a connected Lie group and the following proposition give us a criterion for disjointness from minimal strongly proximal flows of $\mathbb{G}$.

5.1. PROPOSITION: *Let T be a topological group S a closed co-compact subgroup. The following conditions on a minimal flow (T,X) are equivalent.*

(1) (S,X) *is minimal.*

(2) (T,X) *is disjoint from* (T,T/S).

Proof: (1) $\Longrightarrow$ (2): Let W be a closed T-invariant subset of $(T/S)\times X$. Let π_1 denote the projection of W on T/S, then $\pi_1(W) = T/S$. Thus the set $X_o = \{x \in X \mid (S,x) \in W\}$ is not empty. Clearly X_o is closed and S-invariant subset of X. By our assumption $X_o = X$ and it follows that $W = (T/S)\times X$.

(2) $\Longrightarrow$ (1): Let Y be an S-minimal subset of X. Then $T(\{S\}\times Y)$ is a T-invariant subset of $(T/S)\times Y$. We claim that $T(\{S\}\times Y)$ is also closed. Indeed if $\{t_i\}$ is a net in T and $\{y_i\}$ is a net in Y such that

$$\lim t_i(S,y_i) = (tS,x) \in (T/S)\times X$$

exists, then for some $s_i \in S$ $\lim t_i s_i = t$ and we can assume that $\lim s_i^{-1}y_i = y \in Y$ exists. Now

$$x = \lim t_i y_i = \lim(t_i s_i)(s_i^{-1}y_i) = ty,$$

hence $x \in TY$ and $T(\{S\}\times Y)$ is closed. By our assumption $(T/S)\times X$ is T-minimal and thus $T(\{S\}\times Y) = (T/S)\times X \supset \{S\}\times X$. If for $t \in T$, $t(\{S\}\times Y) = \{S\}\times tY$ then $t \in S$ and $tY = Y$, it follows that $X = Y$ and X is S-minimal.//

5.2. THEOREM: *Let* $\mathbb{G}$ *be a connected Lie group,* $(\mathbb{G},X)$ *a minimal flow. Then* $(\mathbb{G},X)$ *is disjoint from* $\pi_S(\mathbb{G})$ *iff* $(\mathbb{H},X)$ *is minimal, where* $\mathbb{H}$ *is a minimal* B-*subgroup of* $\mathbb{G}$.

Proof: This follows from Theorem 3.3. and Proposition 5.1.//

If we take $\mathbb{G}$ to be a semisimple Lie group with finite center $\mathbb{G} = \mathbb{K}\mathbb{A}\mathbb{N}$ an Iwasawa decomposition then $\mathbb{P} = \mathbb{M}\mathbb{A}\mathbb{N}$ is a minimal B-subgroup. Thus $\mathbb{P}$ acts minimally on every minimal $\mathbb{G}$-flow which is disjoint from $\pi_S(\mathbb{G})$, in particular, by III.6.1., $\mathbb{P}$ acts minimally on every G-minimal flow with an invariant measure.

Let Γ be a discrete uniform (i.e. co-compact) subgroup of $\mathbb{G}$. Since Γ is unimodular, 1.6. implies that the flow $(\mathbb{G},\mathbb{G}/\Gamma)$ has an invariant measure and we conclude that $\mathbb{P}$ acts minimally on $\mathbb{G}/\Gamma$. Recently, W.A. Veech proved that moreover $\mathbb{N}$ acts minimally on $\mathbb{G}/\Gamma$, generalizing Hedlund's result about the minimality of the horocyclic flow $(\mathbb{N},\mathbb{G}/\Gamma)$ where $\mathbb{G} = \mathbb{SL}(2,\mathbb{R})$, $\mathbb{N} = \left\{\begin{pmatrix} 1 & b \\ 0 & 1 \end{pmatrix} \mid b \in \mathbb{R}\right\}$ and Γ is a discrete uniform subgroup of $\mathbb{G}$.

Finally we observe that since $(\mathbb{G},\mathbb{G}/\Gamma)$ is a homogeneous flow and since $(\mathbb{G},\mathbb{G}/\Gamma)$ is disjoint from $\mathbb{G}/\mathbb{P}$ we can conclude, by 5.1. that Γ acts minimally on $G/\mathbb{P} \cong \pi_S(\mathbb{G})$.

5.3. THEOREM: *Let* $\mathbb{G}$ *be a connected semisimple Lie group with finite center* Γ *a discrete uniform subgroup and* $\mathbb{B}$ *a* B-*subgroup then* Γ *acts minimally on* $\mathbb{G}/\mathbb{B}$. *In particular if* $\Gamma \subset \mathbb{SL}(n,\mathbb{R})$ *is a uniform and discrete then* Γ *acts minimally on the Grassman variety* $G_{n,n-1}$ *of* m-*dimensional subspaces of* $\mathbb{R}^n$.

This theorem is a generalization of a result of L. Greenberg [4].

IV.6. PROXIMAL FLOWS FOR LIE GROUPS

As was shown in chapter II example 5.5. the group $\mathbb{P} = \left\{\begin{pmatrix} a & b \\ a & a^{-1}\end{pmatrix} \mid a,b \in \mathbb{R}, \ a \neq 0\right\}$ admits a non-trivial minimal proximal flow. However, in this example the action of $\mathbb{P}$ is not continuous with respect to the Lie group topology on $\mathbb{P}$. It is still an open question whether a connected amenable Lie group can admit a non-trivial minimal proximal jointly continuous flow. This question is related to the question whether $\pi_S(\mathbb{G}) = \pi(\mathbb{G})$ for a connected semi-simple Lie group as follows.

6.1. PROPOSITION: Let $\mathbb{G}$ be a connected semisimple Lie group with a finite center. Let $\mathbb{G} = \mathbb{K}\mathbb{S}$ be an Iwasawa decomposition for $\mathbb{G}$ and let $\mathbb{P}$ be the normalizer of $\mathbb{S}$ in $\mathbb{G}$. Then $\pi_S(\mathbb{G}) = \pi(\mathbb{G})$ iff $\pi(\mathbb{P})$ is trivial.

Proof: Suppose $\pi(\mathbb{P})$ is trivial. Since $\mathbb{P}$ is co-compact in $\mathbb{G}$ the flow $(\mathbb{P}, \pi(\mathbb{G}))$ is proximal. By our assumption this implies that $\mathbb{P}$ has a unique fixed point x_o in $\pi(\mathbb{G})$.

Therefore $(\mathbb{G},\pi(\mathbb{G}))$ is a homogeneous flow. Now $\mathbb{P}$ is a maximal B-subgroup (3.2.) and the isotropy group at x_o contains $\mathbb{P}$. Thus $\pi(\mathbb{G})$ is strongly proximal and $\pi(\mathbb{G}) = \pi_S(\mathbb{G})$.

Suppose now that $\pi(\mathbb{P})$ is non-trivial we form the "suspension", $\mathbb{G} \times_{\mathbb{P}} \pi(\mathbb{P})$ which is defined as the quotient of the product space $\mathbb{G} \times \pi(\mathbb{P})$ under the equivalence relation: $(g,x) \sim (g_1,x_1)$ iff there exists $h \in \mathbb{P}$ such that $g_1 = gh$ and $x_1 = hx_1$. The action of $\mathbb{G}$ on the first coordinate of the product $(g(g_1,x) \to (gg_1,x))$ is invariant under this equivalence relation and the induced action makes $(\mathbb{G},\mathbb{G} \times_{\mathbb{P}} \pi(\mathbb{P}))$ a flow. It is easy to see that this flow is minimal and proximal, and clearly it is not isomorphic to $\pi_S(\mathbb{G})$. Thus $\pi(\mathbb{G}) \neq \pi_S(\mathbb{G})$.//

CHAPTER V

μ-HARMONIC FUNCTIONS

In this chapter T is a locally compact topological group with a countable basis for open sets and μ is a fixed probability measure on T. We are interested in solutions of the functional equation

$$h(t) = \int_T h(tt')\,d\mu(t'),$$

which are bounded measurable functions. Such a solution we call a μ-harmonic function. Among the μ-harmonic functions of special interest are the L.U.C. ones.

If (T,X) is a flow and $\nu \in M(X)$ satisfies the convolution equation $\mu * \nu = \nu$ we say that ν is μ-stationary, and ν can be used to obtain L.U.C. μ-harmonic functions as follows; for every continuous function f on X let

$$h(t) = \int_X f\,dt\nu,$$

then h is L.U.C. μ-harmonic on T. Furstenberg has shown [14, 17, and 19] that one can associate with the couple T, μ a flow B and a stationary measure $\nu \in M(B)$ such that every μ-harmonic L.U.C. function arises from B, ν in the above fashion.

We follow Furstenberg [17 and 19] and also Azencott [5] describing this construction and deduce some corollaries of it.

V.1. μ-HARMONIC FUNCTIONS AND AMENABILITY.

As in III.3. let L be the Banach algebra of real valued bounded left uniformly continuous functions on T. Let (T,X) be a flow and ν a measure in $M(X)$. For a function $f \in C(X)$ we define a function $L_\nu f$ on T by:

$$(L_\nu f)(t) = \int_X f \, dt\nu = \int_X f^t \, d\nu.$$

<u>1.1. LEMMA</u>: L_ν <u>is a linear map of</u> $C(X)$ <u>into</u> L. $\|L_\nu\| = 1$, L_ν <u>is positive</u>, <u>sends</u> 1 <u>onto</u> 1 <u>and commutes with</u> T; <u>i.e.</u> $L_\nu(f^t) = (L_\nu f)^t$.

<u>Proof</u>: Clearly L_ν is linear, positive and maps the constant function 1 on X to the constant function 1 on T.

$$\|L_\nu f\| = \sup_t |(L_\nu f)(t)| = \sup_t \left|\int_X f \, dt\nu\right|$$

$$\leq \|f\| \sup_t \int_X dt\nu = \|f\|.$$

Thus $\|L_\nu\| \leq 1$. Since $L_\nu(1) = 1$, $\|L_\nu\| = 1$. For $t \in T$

$$L_\nu(f^t)(s) = \int_X f^t \, ds\nu = \int_X f \, dts\nu =$$

$$= (L_\nu f)(ts) = (L_\nu f)^t(s).$$

Finally if $\{t_i\}$ is a net in T which converges to $t \in T$, then $\|f^{t_i} - f^t\| \to 0$, and

$$\|(L_\nu f)^{t_i} - (L_\nu f)^t\| \leq \|L_\nu\| \|f^{t_i} - f^t\| \to 0.$$

Thus $L_\nu f$ is L.U.C. on T and the lemma is proved.//

For $\nu \in M(X)$ we define $\mu * \nu \in M(X)$ by:

$$(\mu * \nu) f = \int_T \int_X f(tx) \, d\nu(x) \, d\mu(t)$$

$$= \int_T (L_\nu f) \, d\mu \qquad (f \in C(X)).$$

We say that $\nu \in M(X)$ is μ-<u>stationary</u> if $\mu * \nu = \nu$. Denote by $H_\mu = H$ the subspace of L which consists of all L.U.C. μ-harmonic functions.

1.2. PROPOSITION: *The measure* $\nu \in M(X)$ *is* μ-*stationary iff* $L_\nu(C(X)) \subset H_\mu$.

Proof: Let $f \in C(X)$ and put $h = L_\nu f$. $h(t) = \int_X f(tx)\, d\nu(x)$ and

$$\int_T h(tt')\, d\mu(t') = \int_T \int_X f(tt'x)\, d\nu(x)\, d\mu(t') =$$

$$= \int_X f(tx)\, d(\mu * \nu)(x).$$

Thus if $\mu*\nu = \nu$ $h(t) = \int_T h(tt')\, d\mu(t')$ and h is L.U.C. μ-harmonic. Conversely if for every $f \in C(X)$, $h(t) = \int_T h(tt')\, d\mu(t')$ then in particular $h(e) = \int_T h(t')\, d\mu(t')$ i.e. $\int_X f\, d\nu = \int_X f d(\mu*\nu)$ and $\nu = \mu*\nu$. //

We remark that $L_\nu f$ can be defined for every bounded measurable function f on X and the same proof shows that $L_\nu f$ is μ-harmonic iff $\mu*\nu = \nu$.

1.3. LEMMA: *There always exists* $\nu \in M(X)$ *which is* μ-*stationary*.

Proof: Since $\nu \to \mu*\nu$ is an affine continuous map of $M(X)$ into itself, the existence of a μ-stationary ν, is implied by the Markov-Kakutani fixed point theorem. //

1.4. LEMMA: *Let* $\nu \in M(X)$ *be* μ-*stationary, then* ν *is* T-*invariant iff* $L_\nu(C(X)) = \mathbb{R}$.

Proof: For $f \in C(X)$ let $h = L_\nu f$, then $h(t) = h(e)$ for all $t \in T$ iff $\int_X f\, dt\nu = \int_X f\, d\nu$. //

1.5. PROPOSITION: *Let* (T,X) *be a minimal strongly proximal non-trivial flow. Then there exists a* μ-*harmonic non-constant* L.U.C.

function on T.

Proof: By lemma 1.3. there exists $\nu \in M(X)$ which is μ-stationary. Since X is strongle proximal, minimal, and not trivial, ν is not T-invariant. Hence by lemma 1.4. $L_\nu(C(X))$ contains a non-constant function which by proposition 1.2. is L.U.C. μ-harmonic.//

Let S be the subgroup of T generated topologically by the support of μ. Clearly every right S-invariant L.U.C. function on T (i.e. a function which satisfies $f(ts) = f(t)$ for every $t \in T$ and $s \in S$) is μ-harmonic. We say that a topological group T is a *Choquet Deny* (C.D.) *group* if whenever μ is a probability measure on T whose support generates T topologically then $H_\mu = \mathbb{R}$. The classical Choquet Deny theorem asserts that every abelian group is C. D.

By Proposition 1.5. and theorem III.3.1. we have the following:

1.6. THEOREM: *If there exists a probability measure* μ *on* G *such that every* L.U.C. μ-*harmonic function is a constant then* G *is amenable. In particular a* C.D. *group is amenable*.

V.2. CONTRACTIBLE MEASURES

2.1. PROPOSITION: (Azencott) *Let* (T,X) *be a flow and let* $\nu \in M(X)$ *Then the following properties are equivalent*.

(1) *The set* $\{\delta_x \mid x \in X\} = X$ *is contained in the orbit closure of* ν *in* $M(X)$.

(2) L_ν *is an isometry of* $C(X)$ *into* L.

If moreover X *is metrizable then* (1) *and* (2) *are equivalent to*

(3) *There exists a countable dense subset* D *of* X, *and a Borel set* A *such that* A *supports* ν *and for every* $x \in D$ *there exists*

<u>a sequence</u> $\{t_n\}$ <u>of elements of</u> T <u>for which</u> $\lim t_n y = x$ <u>for all</u> $y \in A$.

<u>Proof</u>: (1) ⇒ (2). Since $\|L_\nu\| = 1$, $\|L_\nu(f)\| \leq \|f\|$ for every $f \in C(X)$. Let $x \in X$ be given then there exists a net $\{t_i\}$ in T such that $\lim t_i x = \delta_x$. Hence

$$f(x) = \lim \int_X f \, dt_i\nu = \lim (L_\nu f)(t_i).$$

Thus $|f(x)| \leq \sup_t |(L_\nu f)(t)| = \|L_\nu f\|$ and it follows that $\|f\| \leq \|L_\nu f\|$.

(2) ⇒ (1). Consider the closed convex subset $Q = \overline{co}\,\{t\nu \mid t \in T\}$. If $x \in X$ and $\delta_x \notin Q$, then there exists an affine function F on $M(X)$ such that $F(\delta_x) > \sup F(t\nu) > 0$. Now every affine function on $M(X)$ has the form $\lambda \to \int_X f \, d\lambda$ for some $f \in C(X)$, hence there exists an $f \in C(X)$ such that

$$f(x) > \sup_t \int_X f \, dt\nu = \sup_t (L_\nu f)(t) > 0.$$

This implies $\|f\| > \|L_\nu f\|$ which is a contradiction. Thus we conclude that $\delta_x \in Q$ for every $x \in X$. Since $ex(M(X)) = X$, this means $Q \supset X$ and therefore $Q = M(X)$. Now since $Q = \overline{co}(\text{cls}\,\{t\nu \mid t \in T\})$ it follows that $X \subset \text{cls}\,\{t\nu \mid t \in T\}$.

(3) ⇒ (1). By the dominated convergence theorem, $\lim t_n\nu = \delta_x$ for $x \in D$ and the corresponding sequence t_n. Thus $D \subset \text{cls}\,\{t\nu \mid t \in T\}$ and since D is dense in X, (1) follows.

Finally assuming that X is metric we show that (1) implies (3). Given $x \in X$ we can assume that there exists a sequence $\{t_n\}$ in T such that $\lim t_n\nu = \delta_x$. Let $\{V_m\}$ be a countable basis for the topology at x, and choose a subsequence $\{t_{n_m}\}$ of $\{t_n\}$ such that

$$t_{n_m}\nu(V_m) > 1 - 1/2^m.$$

Let $A_N = \bigcap_{m\geq N} t_{n_m}^{-1}(V_m)$, clearly

$$\nu(A_N) \geq 1 - \sum_{m\geq N} 1/2^m = 1 - 1/2^{N-1}.$$

Denote $A_x = \bigcup_N A_N$ then $\nu(A_x) = 1$ and for every $y \in A_x$, $\lim t_{n_m} y = x$. Given a countable dense set D we now see that $A = \bigcap_{x\in D} A_x$ satisfy (3).//

We say that a measure $\nu \in M(X)$ which satisfies the properties (1) and (2) of proposition 2.1. is a contractible measure.

V.3. BASIC FACTS FROM PROBABILITY THEORY.

Let (Ω,F,P) be a probability triple i.e. Ω is a set, F a σ-algebra of subsets of Ω and P is a σ-additive probability measure defined on F. If (Ω',F',P') is another probability triple and $\phi: \Omega \to \Omega'$ is a map such that $\phi^{-1}(A) \in F$ whenever $A \in F'$ we say that ϕ is measurable or that ϕ is a random variable. If Ω' is a topological space we take F' to be the σ-algebra of Borel sets of Ω'. Given a random variable x, we have a measure xP defined on (Ω',F') by $xP(A) = P(x^{-1}(A))$. We denote this measure by $E^*(x)$ and call it the distribution of x. If x is a real valued random variable we denote the expectation of x by $E(x)$ i.e.,

$$E(x) = \int_\Omega x(\omega)\, dP(\omega) = \int_{\mathbb{R}} \xi\, dE^*(x)(\xi).$$

We denote by $L^p(\Omega,F,P)$ the set of all real valued random variables x, such that

$$E(|x|^p) < \infty \qquad (1 \leq p < \infty).$$

With the norm $E(|x|^p)^{1/p} = \|x\|_p$. L^p is a Banach space.

Let $G \subset F$ be a sub-σ-algebra, then $L^2(\Omega,G,P)$ is a closed subspace of $L^2(\Omega,F,P)$ and the orthogonal projection of $L^2(F)$ on $L^2(G)$ is denoted by $E(\cdot|G)$. For $x \in L^2(G)$, $E(x|G)$ is called the <u>conditional expectation</u> of x with respect to G. We shall use the following properties of the conditional expectation.

(1) $E(z|G)$ is G measurable.

(2) If $z \in L^2(G)$ then $E(z|G) = z$.

(3) For $K \subset G \subset F$ σ-algebras

$$E(z|K) = E(E(z|G)|K).$$

(4) $E(\cdot|G)$ is a contraction on $L^2 \cap L^p$ in the L^p norm for each p.

Since $L^p \cap L^2$ is dense in L^p, (4) implies that $E(\cdot|G)$ can be extended uniquely to L^p. If $\{x_\alpha\}$ is a family of random variables $F(\{x_\alpha\})$ is the least sub-σ-algebra of F with respect to which all the x_α are measurable, we write $E(\cdot|\{x_\alpha\})$ for $E(\cdot|F(\{x_\alpha\}))$.

A family $\{F_\alpha\}$ of subalgebras of F is <u>independent</u> if whenever $A_i \in F_{\alpha_i}$ for distinct $\alpha_1,\cdots,\alpha_p$, $P(A_1 \cap \cdots \cap A_k) = P(A_1)\cdots P(A_k)$. The random variables $\{x_2\}$ are <u>independent</u> if the σ-algebras $F(x_\alpha)$ are independent.

<u>3.1. LEMMA</u>: <u>If</u> $x_1,\cdots,x_n$ <u>are independent random variables and</u> $\phi(x_1,\ldots,x_n) \in L^1(\Omega,F,P)$, <u>then</u>

$$E(\phi(x_1,x_2,\cdots,x_n)|x_1,\cdots,x_{n-1}) =$$

$$= \int \phi(x_1,\cdots,x_{n-1},\xi)\, dE^*(x_n)(\xi).$$

A sequence $w_1,w_2,\cdots$ of real valued random variables form a *martingale* if

$$E(w_n|w_1,\cdots,w_{n-1}) = w_{n-1}; \quad n = 1,2,\cdots.$$

3.2. THEOREM: (Martingale convergence theorem) *Let* $\{w_n\}$ *be a martingale with* $E(|w_n|) < M < \infty$ *for all* n. *Then with probability one,* $\lim w_n$ *exists. If each* $w_n \in L^2(\Omega,F,P)$ *and* $E(|w_m|^2)$ *is bounded then in* L^2 $\lim w_n = w_\infty$ *exists and* $E(w_\infty|w_1,\cdots,w_n) = w_n$.

If $x \in L^1(F)$ and $F_1 \subset F_2 \subset \cdots \subset F$ is a sequence of sub-σ-algebras, then $w_n = E(x|F_n)$ form a martingale. Indeed,

$$E(w_n|w_1,\cdots,w_{n-1}) = E(E(x|F_n)|w_1,\cdots,w_{n-1}) =$$

$$= E(E(E(x|F_n)|F_{n-1})|w_1,\cdots,w_{n-1}) =$$

$$= E(E(x|F_{n-1})|w_1,\cdots,w_{n-1}) = E(w_{n-1}|w_1,\cdots,w_{n-1}) =$$

$$= w_{n-1}.$$

Thus we have the following corollary.

3.3. COROLLARY: *Let* $\{x_i\}_{i=1}^\infty$ *be a sequence of random variables and let* z *be a bounded random variable which is* $F(\{x_i\})$*-measurable then*

$$\lim E(z|x_1,\cdots,x_n) = z \quad \text{almost everywhere.}$$

Proof: If $z \in L^2(\Omega,F\{x_1,\cdots,x_k\},P)$ for some k then clearly

$$E(z|x_1,\cdots,x_n) \to z \quad \text{in} \quad L^2(\Omega,F,P).$$

Since $\bigcup_k L^2(\Omega,F\{x_1,\cdots,x_k\},P)$ is dense in $L^2(\Omega,F(\{x_i\}),P)$ we have this convergence for every $z \in L^2(\Omega,F(\{x_i\}),P)$. Now by theorem 3.2. $\lim (z|x_1,\cdots,x_n)$ exists almost everywhere. Since z is bounded it is in $L^2(\Omega,F(\{x_i\}),P)$ and hence

$$\lim_n E(z|x_1,\cdots,x_n) = z \qquad \text{almost everywhere.}$$

//

V.4. THE POISSON FLOW ASSOCIATED WITH μ.

Let $\{x_n\}$ be a sequence of independent T-valued random variables each with distribution μ. ($E^*(x_n) = \mu$.) Let h be a μ-harmonic function (i.e. $h(t) = E(h(tx_n))$).

4.1. THEOREM: *For a fixed* $t \in T$ *let* $w_0 = h(t)$ *and* $w_n = h(tx_1\cdots x_n)$. *Then with probability one* w_n *converges. Moreover setting* $w_t = \lim w_n$ *we have* $E(w_t) = h(t)$.

Proof: By property (3) (section 3) of the conditional expectation

$$E(w_n|w_1,\cdots,w_{n-1}) = E(E(w_n|x_1,\cdots,x_{n-1})|w_1,\cdots,w_{n-1}).$$

But by lemma 3.1.

$$E(w_n|x_1,\cdots,x_{n-1}) = \int_T h(tx_1\cdots x_{n-1}t')\, dE^*(x_n)(t') =$$

$$= \int_T h(tx_1\cdots x_{n-1}t')\, d\mu(t') = h(tx_1\cdots x_{n-1}) = w_{n-1}.$$

Thus $E(w_n|w_1,\cdots,w_{n-1}) = E(w_{n-1}|w_1,\cdots,w_{n-1}) = w_{n-1}$, and $\{w_n\}$ forms a martingale. Moreover since w_n are bounded, $w_n \in L^2$ and $\|w_n\|_2 \le \|h\|$. Thus by the martingale theorem $\{w_n\}$ converges both in L^2 norm and almost everywhere to an element $w_t \in L^2$ and

$$E(w_t) = \lim E(w_n) = h(t).$$

//

Motivated by this property of harmonic functions we now look at the following collection of functions in L.

$$A = \{f \in L \mid \forall t \in T,\ f(tx_1\cdots x_n) \text{ converges a.e.}\}.$$

It is easy to see that A forms a uniformly closed T-invariant subalgebra of L and by the preceeding theorem $H \subset A$.

Let

$$I = \{f \in L \mid \forall t \in T, \;\; f(tx_1 \cdots x_n) \text{ converges a.e. to } 0\}$$

then I is a closed ideal in A and if Z is the algebra of all limits $z_t = \lim f(tx_1 \cdots x_n)$ where $f \in A$, then $Z \cong A/I$ as algebras.

4.2. LEMMA: Let $f \in A$ and $z_t = \lim f(tx_1 \cdots x_n)$ then $E(z_t) = h(t)$ is a μ-harmonic L.U.C. function (i.e., $h \in H$) and $f-h \in I$.

Proof: By property (3) of the conditional expectation,

$$h(t) = E(z_t) = E(\lim f(tx_1 \cdots x_n)) =$$

$$= \lim E(E(f(tx_1 \cdots x_n) | x_1)) =$$

$$= E(\lim E(f(tx_1 \cdots x_n) | x_1)).$$

By lemma 3.1.

$$\lim E(f(tx_1 \cdots x_n) | x_1) = \lim E_\theta (f(tx_1(\omega) x_2(\theta) \cdots x_n(\theta)))$$

$$= E_\theta \lim f(tx_1(\omega) x_2(\theta) \cdots x_n(\theta)) = E(z_{tx_1}) = h(tx_1),$$

where E_θ denotes expectation with respect to the variable θ on Ω while $\omega \in \Omega$ is held fixed. Hence $h(t) = E(h(tx_1))$ and h is harmonic.

Now

$$h(tx_1 \cdots x_n) = E(z_{tx_1 \cdots x_n}) =$$

$$= \lim E(f(tx_1 \cdots x_n x_{n+1} \cdots x_k) | x_1, \cdots, x_n) =$$

$$= E(z_t | x_1,\cdots,x_n),$$

and since z_t is $\mathcal{F}(x_1,x_2,\cdots)$ measurable $\lim h(tx_1\cdots x_n) = z_t$ by corollary 3.3. Hence $\lim h(tx_1\cdots x_n) = \lim f(tx_1\cdots x_n)$ and $f-h \in \mathcal{I}$. //

4.3. LEMMA: $\mathcal{A} = \mathcal{H} \oplus \mathcal{I}$.

Proof: By lemma 4.2 every $f \in \mathcal{A}$ can be represented as a sum of functions in $\mathcal{H}$ and $\mathcal{I}$, namely $f = h + (f - h)$ where $h(t) = E(z_t)$ and $z_t = \lim f(tx_1\cdots x_n)$. Now if $f \in \mathcal{H} \cap \mathcal{I}$ then $z_t = 0$ and by theorem 4.1. $f(t) = E(z_t) = 0$ for every $t \in T$. //

We now have $\mathcal{Z} \cong \mathcal{A}/\mathcal{I} \cong \mathcal{H}$ and we can use this isomorphism to define a multiplication on the Banach space $\mathcal{H}$ which will make it a Banach algebra isomorphic to $\mathcal{Z}$ or $\mathcal{A}/\mathcal{I}$. We conclude that there exists a compact Hausdorff space B such that $\mathcal{H}$ is isometrically isomorphic to $C(B)$ as Banach algebras. Moreover the action of T on $\mathcal{A}$ $(f \to f^t)$ induces an action of T on B which makes it a flow under T and the isomorphism of $C(B)$ with $\mathcal{H}$ commutes with the action of T on these spaces.

The maps $h \to h(e)$ of $\mathcal{H}$ into $\mathbb{R}$ is a continuous linear functional, and by the Riesz representation theorem there exists a probability measure ν on B such that if $\tilde{h}$ is the continuous function on B which corresponds to h, then

$$h(e) = \int_B \tilde{h}(\xi)\, d\nu(\xi).$$

Moreover for $t \in T$, h^t corresponds to $\tilde{h}^t$ and thus

$$h(t) = h^t(e) = \int_B \tilde{h}^t(\xi)\, d\nu(\xi) = \int_B \tilde{h}(\xi)\, d(t\nu)(\xi) =$$

$$= (L_\nu \tilde{h})(t).$$

Thus L_ν sends $C(B)$ isometrically onto H. By proposition 2.1. ν is contratible and by proposition 1.2. ν is μ-stationary. We say that (B,ν) is the Poisson flow, and Poisson measure associated with μ.

4.4. THEOREM: *Let* (T,X) *be a flow* $\lambda \in M(X)$. *The following conditions are equivalent.*

(1) L_λ *is an isometry of* $C(X)$ *into* H.

(2) λ *is* μ-*stationary and contractible.*

(3) *There exists a flow homomorphism into,* $\phi: B \to M(X)$ *such that* $\phi(B) \supset X$ *and* $\tilde{\phi}(\nu) = \lambda$ *where* $\tilde{\phi}$ *is the unique affine extension of* ϕ *to* $M(B)$.

Proof: The equivalence of (1) and (2) was proved in propositions 1.2. and 2.1.

Suppose that (1) holds and let $\tilde{\phi} = (L_\nu^{-1}L_\lambda)^*|M(B)$ be the restriction of the adjoint of $L_\nu^{-1}L_\lambda$ to $M(B)$. $\tilde{\phi}$ is a (weak $*$) continuous affine map which commutes with T. If $f \in C(X)$, denote $h = L_\lambda f$ and let $\tilde{h} = L_\nu^{-1}h$, then

$$\int_X f\, d\tilde{\phi}(\nu) = \int_B (L_\nu^{-1}L_\lambda)f\, d\nu = \int_B L_\nu^{-1}h\, d\nu =$$

$$= \int_B \tilde{h}\, d\nu = h(e) = (L_\lambda f)(e) = \int_X f\, d\lambda.$$

Thus $\tilde{\phi}(\nu) = \lambda$. Since (1) and (2) are equivalent, λ is contractible. Therefore, X is contained in the image of $\tilde{\phi}$. Finally since the set of extreme points in $M(X)$ is X and that of $M(B)$ is B, it follows that $\tilde{\phi}(B) \supset X$. Thus $\tilde{\phi}|B = \phi$ is the required map.

Conversely if ϕ is a given homomorphism of B into $M(X)$ with $\phi(B) \supset X$. Then ϕ can be extended to a homomorphism of $M(B)$ into $M(M(X))$ and composing this extension with the barycenter map of

$M(M(X))$ onto $M(X)$, we get the unique extension of ϕ which maps $M(B)$ onto $M(X)$. By our assumption $\tilde{\phi}(\nu) = \lambda$ and since ν is contractible and μ-stationary so is λ.//

We observe that in the case that (T,B) is a minimal flow we must have $\phi(B) = X$ and in the case (B,ν) is a universal couple in the usual sense.

4.5. COROLLARY: *The universal minimal strongly proximal flow is a factor of every minimal set of* B.

Proof: Every probability measure on $\pi_S(T)$ is contractible and some of these are μ-stationary.//

V.5. C.D.-GROUPS

We recall that a C.D.-group is a topological group T for which $H_\mu = \mathbb{R}$ whenever μ is a probability measure on T such that the support of μ generates T topologically. (i.e. the subgroup of T generated by the support of μ is dense in T.) The following is a slight extension of a theorem of Choquet Deny [6].

5.1. THEOREM: *Let* T *be a topological group* S *its center, and suppose that* T/S *is compact then* T *is a* C.D.-*group. In particular abelian groups and compact groups are* C.D.-*groups*.

Proof: Let μ be a probability measure on T whose support generates T topologically, and let (B,ν) be the associated Poisson flow and Poisson measure. Let

$$Q = \{\lambda \in M(B) \mid \mu * \lambda = \lambda\},$$

then Q is a closed convex non-empty $(\nu \in Q)$ S-invariant subset of $M(B)$. Let x be an arbitrary point of B, then there exists a net $\{t_i\}$ in T such that $\lim t_i\nu = \delta_x$. Since T/S is compact, we can assume that in T/S $\lim t_i S = tS$ exists. This implies that for a net $\{s_i\}$ in S $\lim t_i s_i = t$. Again we can assume that $\lim s_i^{-1}\nu = \lambda$ exists and thus

$$\delta_x = \lim t_i \nu = \lim (t_i s_i)(s_i^{-1}\nu) = t\lambda .$$

Therefore $\lambda = t^{-1}\delta_x = \delta_{t^{-1}x}$ and since Q is S-invariant $\delta_{t^{-1}x} \in Q$. Now the equality $\mu * \delta_{t^{-1}x} = \delta_{t^{-1}x}$ clearly implies that the point $t^{-1}x$ is fixed under the support of μ and hence $t^{-1}x$ is a fixed point for T. We conclude that x is a fixed point for T and since x was arbitrary, the action of T on B and hence also on $M(B)$ is trivial. Since ν is contractible, this implies that B is a trivial one point flow and thus H_μ which is isomorphic to $C(B)$ consists of the constant function.
//

The following theorem appears in [19] with an indication of a proof. We give here an alternative proof.

5.2. THEOREM (Furstenberg): *Let* T *be a nilpotent group* μ *a probability measure on* T *whose support generates* T *topologically. If for every* $t \in T$ *there exists an* n *such that* μ^n *and* $t\mu^n$ *are not mutually singular then* $H_\mu = \mathbb{R}$. ($\mu^n = \mu * \cdots * \mu$ n *times*).

Proof: Let t be an element of the center of T and let n be such that $t\mu^n \perp \mu^n$. Let (B,ν) be the Poisson pair for (T,μ), then $\mu * \nu = \nu$ implies $\mu^n * \nu = \nu$ and $t\nu = t\mu^n * \nu$ is not singular to $\nu = \mu^n * \nu$. Since ν is a contractible measure we can assume (by passing to an arbitrary metric factor of (B,ν)) that there exists a countable dense set $D \subset B$ and a Borel set $A \subset B$ such that A supports ν and such that for every $x \in D$ there exists a sequence $\{t_i\}$ of elements of T for which $\lim t_i y = x$ for all $y \in A$ (Proposition 2.1.).

Since $t\nu \perp \nu$ there exists $z \in A$ such that also $tz \in A$. If

$x \in D$ then let $\{t_i\}$ be as above; we have $x = \lim t_i z = \lim t_i tz = t \lim t_i z = tx$. Thus $tx = x$ for all $x \in D$ hence for all $x \in B$. By induction on a central series for T we conclude that T acts trivially on B. This implies that B is a point and $H_\mu = \mathbb{R}$. //

5.3. COROLLARY: A discrete countable nilpotent group is a C.D.-group.

Proof: Let T be a discrete nilpotent group and let μ be a probability measure on T whose support generates T as a group. Let S be the subsemigroup generated by the support of μ. It is easy to see that $S = \cup \operatorname{Supp}(\mu^n)$. By [5, lemma IV.11] SS^{-1} is a subgroup of T. (Lemma IV.11. of [5] states that whenever S is a subsemigroup of a nilpotent group then SS^{-1} is a group. The proof is purely algebraic and easy.)

Thus our assumption on μ implies $SS^{-1} = T$ and if $t \in T$ then $tS \cap S \neq \emptyset$. Therefore for some n, $t\mu^n \not\perp \mu^n$ (recall that T is discrete) and our corollary follows from theorem 5.2. //

In [24, p. 372] Y. Guivarch proves that a nilpotent group of degree 2 as well as a compact abelian extension of an abelian group are C.D.-groups. It also follows from his work as also from R. Azencott's that a discrete nilpotent group is a C.D.-group. It is an open question whether a nilpotent group is a C.D. group.

A related notion to that of a C.D.-group is the notion of a Liouville group (see [24, p. 368].) A probability measure μ on a locally compact topological group is called étallé if for some n, $\mu^n = \mu * \mu * \cdots * \mu$ is not singular with respect to the Haar measure on T. T is called a Liouville group if whenever μ is étallé and the support of μ generates T topologically then $H_\mu = \mathbb{R}$. Azencott showed that a connected Lie group which is a compact extension of a solvable group of type R, is a Liouville group [5]. (A connected

Lie group G is of type R is the eigenvalues of Ad(g) are of absolute value one for each $g \in \mathbb{G}$.). This was generalized by Guivarch.

CHAPTER VI

HARMONIC FUNCTIONS ON A SYMMETRIC SPACE

The material in this chapter is due to Furstenberg and the main sources for it are [14] and [17]. In all of this chapter $\mathbb{G}$ is a connected semisimple Lie group with a finite center and $\mathbb{K}$ is a maximal compact subgroup. Let m be the unique $\mathbb{K}$-invariant measure on $\pi_S(\mathbb{G})$. Our first goal is to prove that the map ϕ: $g\mathbb{K} \to gm$ of $D = \mathbb{G}/\mathbb{K}$ into $M(\pi_S(\mathbb{G}))$ is a homeomorphism (Theorem 2.1). As a preliminaty result we show in section one that if $\mathbb{G}$ is simple then $\mathbb{K}$ is a maximal closed subgroup of $\mathbb{G}$ from which it follows easily that ϕ is one-to-one. This approach was suggested by R. Lipsman. The space $\overline{\phi(D)}$ is the Furstenberg compactification of D and clearly $\pi_S(\mathbb{G})$ is a subset of this compactification.

In section three we define a class of measures on $\mathbb{G}$, the spherical measures, and we find several characterizations for functions which are μ-harmonic for every spherical measure μ on $\mathbb{G}$ (Theorem 3.1.). (If f is μ-harmonic for a spherical measure μ, then f is necessarily right $\mathbb{K}$-invariant so that f is defined on $D = \mathbb{G}/\mathbb{K}$.)

Chapter III of [14] is devoted to the proof of a lemma (lemma 3.2. here) which states that a left $\mathbb{K}$-invariant μ-harmonic function is a constant whenever μ is an absolutely continuous measure on $\mathbb{G}$. We do not include the proof of this deep lemma but we conclude from it that if f is μ-harmonic for one spherical measure it is harmonic for all of them. In section 4 we indicate how this fact can be used to identify the, common, Poisson space and measure (B,ν) for all the spherical measures on $\mathbb{G}$ as the maximal boundary $\pi_S(\mathbb{G})$ and the measure m respectively. In section 5 we show that the theory of harmonic functions on D reduces to the classical theory of harmonic functions in the unit disk when $\mathbb{G} = \mathbb{SL}(2,\mathbb{R})$.

1. FOR $\mathbb{G}$ SIMPLE, $\mathbb{K}$ IS A MAXIMAL CLOSED SUBGROUP OF $\mathbb{G}$.

1.1. LEMMA: *Let g be a semisimple Lie algebra $g = t \oplus p$ a Cartan decomposition. If a and b are subspaces of p such that $\mathrm{ad}(t)b \subset b$ and $B(a,b) = 0$ (where B is the Killing form). Then $[a,b] = 0$.*

Proof: Let $\gamma = [\alpha,\beta]$ where $\alpha \in a$ and $\beta \in b$, then $\gamma \in t$ and hence

$$\delta = [\beta,[\alpha,\beta]] \in [b,t] \subset b.$$

Now

$$B(\gamma,\gamma) = B([\alpha,\beta],\gamma) =$$

$$= B([\beta,\gamma],\alpha) =$$

$$= B(\delta,\alpha) = 0.$$

On the other hand by IV.1.4. B is negative definite on t and thus $B(\gamma,\gamma) = 0$ implies $\gamma = 0$, i.e. $[a,b] = 0$. //

1.2. LEMMA: *In the notations of lemma 1.1. suppose that g is simple then $\mathrm{ad}(t)$ acts irreducibly on p.*

Proof: Let $g_0 = p \oplus [p,p]$; this is a subspace of g and since p is $\mathrm{ad}(t)$ invariant it is easy to see, by the Jacobi identity, that g_0 is an ideal of g. Since g is simple $g = g_0$ and it follows that $[p,p] = t$.

Let $p_1 \subset p$ be an $\mathrm{ad}(t)$ invariant subspace of p, and let $g_1 = [p_1,p_1] \oplus p_1$. Let p_2 be the orthogonal complement of p_1 in p with respect to B; then $p = p_1 \oplus p_2$ and by lemma 1.1. $[p_1,p_2] = 0$. Now

$$g = t \oplus p = [p,p] \oplus p = [p_1 \oplus p_2, p_1 \oplus p_2] \oplus p_1 \oplus p_2$$

and it follows from the $ad(t)$ invariance of p_1 and from the Jacobi identity that $[g,g_1] \subset g_1$ i.e. g_1 is an ideal of g. Since g is simple $g_1 = 0$ or $g = g_1$ and it follows that $p_1 = 0$ or that $p = p_1$ respectively. The proof is completed.

//

<u>1.3. THEOREM</u>: <u>Let $\mathbb{G}$ be a connected simple Lie group with a finite center, $\mathbb{K}$ a maximal compact subgroup, then $\mathbb{K}$ is a maximal closed subgroup of $\mathbb{G}$.</u>

<u>Proof</u>: Let $\mathbb{L}$ be a closed subgroup of $\mathbb{G}$ containing $\mathbb{K}$. Let t, ℓ and g be the Lie algebras of $\mathbb{K}$, $\mathbb{L}$ and $\mathbb{G}$ respectively and let $g = t \oplus p$ be a Cartan decomposition of g. If we denote by p_1 the projection of ℓ on p and recall that both ℓ and p are $ad(t)$ invariant then it is clear that p_1 is also $ad(t)$ invariant. By lemma 1.2. $p_1 = p$ or $p_1 = 0$. In the first case $\ell = g$ and $\mathbb{L} = \mathbb{G}$. In the second case $\mathbb{L}^o$ the identity component of $\mathbb{L}$ is equal to $\mathbb{K}$, and $\mathbb{L}/\mathbb{K}$ is discrete. The proof of the theorem will be completed by the following lemma.

<u>1.4. LEMMA</u>: <u>Let $\mathbb{G}$ be a connected semisimple Lie group with finite center. Let $\mathbb{K}$ be a maximal compact subgroup of $\mathbb{G}$ and let $\mathbb{L}$ be a closed subgroup of $\mathbb{G}$ containing $\mathbb{K}$ such that $\mathbb{L}/\mathbb{K}$ is discrete then $\mathbb{K} = \mathbb{L}$.</u>

<u>Proof</u>: Let $\phi: \mathbb{G} \to \mathbb{G}/\mathbb{K}$ be the natural map and denote $x_o = \phi(\mathbb{K})$. $\phi(\mathbb{L})$ is a discrete subset of $\mathbb{G}/\mathbb{K}$ and since $\mathbb{K} \subseteq \mathbb{L}$ the action of $\mathbb{K}$ on $\mathbb{G}/\mathbb{K}$ (on the left) leaves $\phi(\mathbb{L})$ invariant. If there is an $\ell \in \mathbb{L}$ such that $\ell \notin \mathbb{K}$ we denote $\phi(\ell) = x$ and by the discreteness of $\phi(\mathbb{L})$, $\mathbb{K}x = x$. Let $g = t \oplus p$ be a Cartan decomposition. Then the map $\psi: \mathbb{K} \times p \to \mathbb{G}$, $\psi(k,\alpha) = k\exp(\alpha)$ is an analytic isomorphism and hence every element of $\mathbb{G}/\mathbb{K}$ has a unique representation as $\exp(\alpha)x_o$

where $\alpha \in p$. (IV.1.4.)

Let $x = \exp(\alpha)x_o$ and $k \in \mathbb{K}$ then

$$\exp(\alpha)x_o = x = kx = k\exp(\alpha)x_o = k\exp(\alpha)k^{-1}x_o = \exp(Ad(k)\alpha)x_o.$$

But p is invariant under $Ad(k)$ (IV.1.4.), hence $Ad(k)\alpha \in p$ and it follows from the uniqueness of the representation, that $Ad(k)\alpha = \alpha$. This implies $[t,\alpha] = 0$ and by lemma 1.4 $\alpha \in t$ i.e. $\alpha = 0$. This is a contradiction to our assumption that $x = \phi(\ell) = \exp(\alpha)x_o$ is not equal to x_o. We conclude that $\phi(\mathbb{L}) = x_o$ and thus $\mathbb{L} \subset \mathbb{K}$.
//

This completes the proof of theorem 1.3.

VI.2. THE FURSTENBERG COMPACTIFICATION OF A SYMMETRIC RIEMANIAN SPACE

Let $\mathbb{G}$ be a connected semisimple Lie group with a finite center, $\mathbb{K}$ a maximal compact subgroup. The space $\mathbb{G}/\mathbb{K}$ has a natural structure as a riemanian space and with this structure D is a symmetric space i.e. each $x \in D$ is an isolated point of an involutive isometry of D. (This involutive isometry is unique and is given by the geodesic symmetry around x. [26]).

Let $\mathbb{G} = \mathbb{K}\mathbb{S}$ be an Iwasawa decomposition of $\mathbb{G}$ and $\mathbb{P} = \mathbb{M}\mathbb{S}$ be a minimal B-subgroup of $\mathbb{G}$ (see IV.3.). Then $\pi_S(\mathbb{G})$ is isomorphic to $\mathbb{G}/\mathbb{P}$. Since $\mathbb{K}$ is transitive on $\mathbb{G}/\mathbb{P}$ there exists a unique $\mathbb{K}$-invariant probability measure $m \in M(\pi_S(\mathbb{G}))$. Consider the map of $D = \mathbb{G}/\mathbb{K}$ into $M(\pi_S(\mathbb{G}))$ defined by:

$$g\mathbb{K} \xrightarrow{\phi} gm \qquad (g \in \mathbb{G}).$$

<u>2.1. THEOREM</u>: <u>The map</u> ϕ <u>is a homeomorphism of</u> D <u>into</u> $M(\pi_S(\mathbb{G}))$.

<u>Proof</u>: First assume that $\mathbb{G}$ is simple and set $\mathbb{L} = \{g \in \mathbb{G} : gm = m\}$; then $\mathbb{L}$ is a closed subgroup of $\mathbb{G}$ containing $\mathbb{K}$. By theorem 1.3. either $\mathbb{L} = \mathbb{K}$ or $\mathbb{L} = \mathbb{G}$. If $\mathbb{L} = \mathbb{G}$, then m would be a $\mathbb{G}$-invari-

ant measure on $\mathbb{G}/\mathbb{P} = \pi_S(\mathbb{G})$. Therefore, ϕ is one-to-one.

If $\mathbb{G}$ is not simple we first observe that since $\mathbb{K}$ contains the center of $\mathbb{G}$, we can assume that $\mathbb{G}$ has a trivial center. In that case $\mathbb{G} = \mathbb{G}_1 \times \cdots \times \mathbb{G}_n$ is a direct product of simple groups and $\mathbb{K} = \mathbb{K}_1 \times \cdots \times \mathbb{K}_n$, where $\mathbb{K}_i$ is a maximal compact subgroup of $\mathbb{G}_i$. The flow $(\mathbb{G}_i, \pi_S(\mathbb{G}_i))$ is a factor of the flow $(\mathbb{G}, \pi_S(\mathbb{G}))$ and clearly m is mapped onto m_i, the unique $\mathbb{K}_i$-invariant probability measure on $\pi_S(\mathbb{G}_i)$. Now if $gm = m$ for $g \in \mathbb{G}$ then $g_i m_i = m_i$, where g_i is the ith component of g. Hence $g_i \in \mathbb{K}_i$ and it follows that $g \in \mathbb{K}$. The proof that ϕ is 1-1 is complete.

In order to show that ϕ is a homeomorphism one should show that the convergence of a sequence of the form $g_i m$ to a measure g^m $(g_i, g \in \mathbb{G})$, implies the convergence $g_i\mathbb{K} \to g\mathbb{K}$ in D. Now as in the proof of [33, theorem 4] each g_i can be written as $k_i a_i k_i'$ for $k_i, k_i' \in \mathbb{K}$ and $a_i \in \mathbb{A}$, and $g_i m = k_i a_i m$. Thus it is enough to show that if the sequence $\{a_i\}$ is not bounded then the measure $\lim a_i m$ is supported by a submanifold of $\pi_S(\mathbb{G})$ of a strictly lower dimension than that of $\pi_S(\mathbb{G})$. This can be done in a way similar to the proof of theorem 1 of [33]. We do not go into further details of the proof. //

We now identify D with its image $\phi(D)$ in $M(\pi_S(\mathbb{G}))$. Taking the closure of D in $M(\pi_S(\mathbb{G}))$ we obtain a compactification of D. Since $\pi_S(\mathbb{G})$ is strongly proximal and minimal, every point of $\pi_S(\mathbb{G})$ is an accumulation point of the set $\mathbb{G}m = D$. Thus $\pi_S(\mathbb{G})$ is a part of the boundary of D in $M(\pi_S(\mathbb{G}))$. It can be shown that $\bar{D}$ consists of finitely many $\mathbb{G}$-orbits of which only $\pi_S(\mathbb{G})$ is compact. It was shown by C.C. Moore [33] that this compactification of D (the <u>Furstenberg compactification</u>) coincide with one of the Satake compactifications.

VI.3. HARMONIC FUNCTIONS ON D.

Throughout the rest of this chapter we use the notation, $X = \pi_S(\mathbb{G})$ and $M = M(X)$. Thus $X \subset \bar{D} \subset M$. We recall that m is the unique $\mathbb{K}$-invariant measure in M. For a function f on D we denote by $\tilde{f}$ the lifted function on $\mathbb{G}$; i.e., $\tilde{f}(g) = f(gm)$.

If T is a locally compact topological group we denote by $\int_T \cdot\, dt$ integration with respect to Haar measure on T and we write $L^p(T)$ for $L^p(T,dt)$. Let $f \in L^\infty(\mathbb{G})$ and $g \in \mathbb{G}$ we define ${}^g f(g_1) = f(g_1 g)$. The linear operator $R_g: f \to {}^g f$ is an isometry of $L^\infty(\mathbb{G})$ onto itself and we refer to the representation $g \to R_g$ of $\mathbb{G}$ on $L^\infty(\mathbb{G})$ as the <u>regular representation</u> of $\mathbb{G}$.

For $f \in L^\infty(\mathbb{G})$ let $V(f)$ denote the smallest compact convex subset of $L^\infty(\mathbb{G})$ which is invariant under the regular representation of $\mathbb{G}$ and contains f. (Compact means here compact in the weak $*$ topology induced on $L^\infty(\mathbb{G})$ by $L^1(\mathbb{G})$.).

We say that a probability measure μ on $\mathbb{G}$ is <u>spherical</u> if it is absolutely continuous with respect to Haar measure on $\mathbb{G}$ and if in addition $k\mu = \mu k = \mu$ for every $k \in \mathbb{K}$. In other words there exists $\psi \in L^1(\mathbb{G})$ such that $\psi \geq 0$, $\int_{\mathbb{G}} \psi(g)\,dg = 1$, $\psi(kg) = \psi(gk) = \psi(g)$ for every $k \in \mathbb{K}$ and $g \in \mathbb{G}$ and $\int_{\mathbb{G}} f\,d\mu = \int_{\mathbb{G}} f\psi\,dg$.

3.1. THEOREM: <u>Let f be a continuous bounded function on D. The following conditions on f are equivalent.</u>

(1) <u>There exists a function</u> $\phi \in L^\infty(X,m)$ <u>such that</u>

$$\tilde{f}(g) = \int_X \phi(gx)\,dm(x) = (L_m\phi)(g).$$

(2) $\tilde{f}$ <u>is μ-harmonic for every spherical measure</u> μ <u>on</u> $\mathbb{G}$.

(3) $\tilde{f}(g) = \int_{\mathbb{K}} \tilde{f}(gkg')\,dk$ <u>for every</u> g <u>and</u> $g' \in \mathbb{G}$.

(4) <u>The regular representation of</u> $\mathbb{G}$ <u>on</u> $V(\tilde{f})$ <u>is irreducible</u> (see

III.2.).

We say that a continuous bounded function f on $\mathbb{G}$ is <u>harmonic</u> if it satisfies properties (1)-(4) above. Note that if f is harmonic then $f(gk) = f(g)$ for all $g \in \mathbb{G}$ and $k \in \mathbb{K}$ so that f is a lift of a function on D which we also call harmonic.

<u>Proof</u>: (1) => (2) Let μ be a spherical measure then $k(\mu * m) = (k\mu)*m = \mu*m$. Hence $\mu*m$ is $\mathbb{K}$-invariant and by the uniqueness of m, $\mu*m = m$. Now the operator L_m can be defined on $L^\infty(X,m)$ as it was defined on $C(X)$ namely:

$$(L_m\psi)(g) = \int_X \psi \, d(gm) \qquad (\psi \in L^\infty(X,m)),$$

and the proof of proposition V.1.2. shows that $L_m\psi$ is μ-harmonic for every ψ. In particular $\tilde{f} = L_m\psi$ is μ-harmonic.

(2) => (3) For a fixed $g \in \mathbb{G}$ let F be defined by:

$$F(g') = \int_{\mathbb{K}} \tilde{f}(gkg') \, dk.$$

F is continuous bounded function of g'. We consider F as an element of $L^\infty(\mathbb{G})$ and let $\phi \in L^1(\mathbb{G})$. Suppose that $\phi \ge 0$ and that $\int_{\mathbb{G}} \phi(g) \, dg = 1$. Then

$$\int_{\mathbb{G}} F(g')\phi(g') \, dg' = \int_{\mathbb{G}} \int_{\mathbb{K}} \tilde{f}(gkg')\phi(g') \, dk \, dg'.$$

Since $\tilde{f}$ is right $\mathbb{K}$-invariant this is equal to

$$\int_{\mathbb{G}} \int_{\mathbb{K}} \int_{\mathbb{K}} \tilde{f}(gkg'k')\phi(g') \, dk \, dk' \, dg'.$$

Since dg is left and right $\mathbb{G}$-invariant the last integral is equal to

$$\int_{\mathbb{G}} \int_{\mathbb{K}} \int_{\mathbb{K}} f(gg')\phi(k^{-1}g'k'^{-1}) \, dk \, dk' \, dg' =$$

$$\int_{\mathbb{G}} f(gg')\psi(g')\, dg'$$

where

$$\psi(g') = \int_{\mathbb{K}} \int_{\mathbb{K}} \phi(kg'k')\, dk\, dk'.$$

If we let $d\mu = \psi dg$ then μ is spherical and by (2)

$$\int_{\mathbb{G}} f(gg')\, d\mu = \tilde{f}(g).$$

Thus $\int_{\mathbb{G}} F(g')\phi(g')\, dg' = \tilde{f}(g)$ for every ϕ in $L^1(\mathbb{G})$ and we conclude that as elements of $L^\infty(\mathbb{G})$, $F(g')$ and the constant function $\tilde{f}(g)$ are equal. Since both functions are continuous there are equal everywhere.

(3) => (4) Let

$$Q_f = \{F \in L^\infty(\mathbb{G}) \mid \|F\|_\infty \le \|\tilde{f}\|_\infty \text{ and for almost all } g'$$
$$\tilde{f}(g) = \int_{\mathbb{K}} F(gkg')\, dk \text{ as elements of } L^\infty(\mathbb{G})\}.$$

By (3) $\bar{f} \in Q_f$ and it is clear that Q_f is a convex subset of $L^\infty(\mathbb{G})$ which is invariant under the regular representation. We show next that Q_f is weak * closed. Indeed $F \in Q_f$ iff $\|F\|_\infty \le \|\tilde{f}\|_\infty$ and for almost all g'

$$\int_{\mathbb{G}} \tilde{f}(g)\phi(g)\, dg = \int_{\mathbb{G}} \int_{\mathbb{K}} F(gkg')\phi(g)\, dk\, dg$$

for every $\phi \in L^1(\mathbb{G})$.

Now

$$\int_{\mathbb{G}} \int_{\mathbb{K}} F(gkg')\phi(g)\, dk\, dg = \int_{\mathbb{K}} \int_{\mathbb{G}} F(gkg')\ {}^{(kg')^{-1}}\phi(gkg')\, dg\, dk =$$

$$\int_{\mathbb{K}} \int_{\mathbb{G}} F(g)\phi(gg'^{-1}k^{-1})\, dg\, dk = \int_{\mathbb{G}} F(g) \int_{\mathbb{K}} \phi(gg'^{-1}k^{-1})\, dk\, dg =$$

$$\int_G F(g)\psi(gg')\,dg \qquad \text{where} \qquad \psi(gg') = \int_{\mathbb{K}} \phi(gg'^{-1}k)\,dk.$$

Since for almost all g', $\psi(gg') \in L'(\mathbb{G})$ it follows that if F_i converges weak * to F and $F_i \in Q_f$ then $F \in Q_f$; i.e., Q_f is weak * closed. Since Q_f is bounded, it follows that Q_f is weak * compact. Since $\tilde{f} \in Q_f$, $Q_f \supseteq V(\tilde{f})$. Therefore if $F \in V(\tilde{f})$ then for almost all g'

$$\tilde{f}(g) = \int_{\mathbb{K}} F(gkg')\,dk.$$

If ψ is an element of $L^1(\mathbb{G})$ such that $\psi \geq 0$ and $\int_{\mathbb{G}} \psi(g)\,dg = 1$ then in $L^\infty(\mathbb{G})$

$$\int_{\mathbb{G}} \int_{\mathbb{K}} F(gkg')\psi(g')\,dk\,dg' = \tilde{f}(g).$$

Now clearly the function of the left side of the above equality belongs to $V(F)$, (as a limit of convex combinations of translates of $F(g)$). Thus $\tilde{f} \in V(f)$ whenever $F \in V(\tilde{f})$; i.e., $V(\tilde{F})$ is irreducible.

(4) => (1) We now assume that $V(\check{f})$ is irreducible. By the universality of $M(X)$ (proposition III.2.4.) there exists a unique affine homomorphism $M(X) \xrightarrow{\tau} V(\tilde{f})$. Let $x_0 \in X$ and put $\tau(x_0) = \Phi_0$, then $\mathbb{H}_0 = \{h \in \mathbb{G} \mid hx_0 = x_0\}$ is a B-subgroup of $\mathbb{G}$ and for $h \in \mathbb{H}_0$

$$^h\Phi_0 = \tau(hx_0) = \tau(x_0) = \Phi_0.$$

Thus $\Phi_0(gh) = \Phi_0(g)$ almost everywhere in g for each $h \in \mathbb{H}_0$. We show next that Φ_0 can be replaced by a function Φ which represents the same element of $L^\infty(\mathbb{G})$ and such that $\Phi(gh) = \Phi(g)$ almost everywhere in g for <u>all</u> $h \in \mathbb{H}_0$. Let θ be a continuous non-zero function on $\mathbb{H}_0$ with a compact support and such that $\int \theta(h)\,dh = 1$. Let

$$\Phi(g) = \int_{\mathbb{H}_0} \Phi_0(gh)\theta(h)\,dh\,.$$

If $\psi \in L^1(\mathbb{G})$ then

$$\int_{\mathbb{G}} \Phi(g)\psi(g)\, dg = \int_{\mathbb{G}} \int_{\mathbb{H}_o} \Phi_o(gh)\theta(h)\psi(g)\, dh\, dg =$$

$$\int_{\mathbb{H}_o} \left(\int_{\mathbb{G}} \Phi_o(gh)\psi(g)\, dg\right) \theta(h)\, dh.$$

But for a fixed h, $\Phi_o(gh) = \Phi_o(g)$ g a.e. hence

$$\int_{\mathbb{G}} \Phi(g)\psi(g)\, dg = \int_{\mathbb{G}} \Phi_o(g)\psi(g)\, dg.$$

So that $\Phi(g) = \Phi_o(g)$ as elements of $L^\infty(\mathbb{G})$. Now for a fixed g, $\Phi(gh)$ is a continuous function of h hence if H is a dense countable subset of $\mathbb{H}_o$ and if $\Phi(gh) = \Phi(g)$ for every $h \in H$ then $\Phi(gh) = \Phi(g)$ for every $h \in \mathbb{H}_o$. It follows that $\Phi(gh) = \Phi(g)$ g a.e. for all $h \in \mathbb{H}_o$.

We can now define a function ϕ on X by $\phi(gx_o) = \Phi(g)$. ϕ is well defined measurable bounded function on X. For $x \in X$ there exists $g \in \mathbb{G}$ with $x = gx_o$, hence if $g' \in \mathbb{G}$

$$\tau(x)(g') = \tau(gx_o)(g') = (g\tau(x_o))(g') = {}^g\Phi(g')$$

$$= \Phi(g'g) = \phi(g'gx_o) = \phi(g'x).$$

Now $\tilde{f}$ is a $\mathbb{K}$-invariant point of $V(\tilde{f})$ and m is a unique $\mathbb{K}$-invariant point of M, hence $\tau(m) = \tilde{f}$.

We write the weak $*$ integral representation $m = \int_X \delta_x\, dm(x)$ and hence for every $g \in \mathbb{G}$

$$\tilde{f}(g) = \tau(m)(g) = [\tau(\int_X \delta_x\, dm(x))](g) =$$

$$= \left(\int_X \tau(x)\, dm(x)\right)(g) = \int_X \tau(x)(g)\, dm(x) =$$

$$= \int_X \phi(gx)\, dm(x) = (L_m\phi)(g). \quad //$$

It turns out that for a function f to be harmonic it is enough that f is μ-harmonic for some spherical measure μ. This is a result of the following deep lemma of Furstenberg. (For the proof see chapter III of [14]).

3.2. LEMMA: Let μ be an absolutely continuous probability measure on $\mathbb{G}$, let $f(g)$ be a μ-harmonic function on $\mathbb{G}$ such that $f(kg) = f(g)$ for every $k \in \mathbb{K}$ and $g \in \mathbb{G}$. Then $f(g)$ is a constant.

3.3. THEOREM: Let f be a continuous bounded μ-harmonic function for some spherical measure μ. Then f is harmonic. Thus the Banach spaces H_μ of L.U.C. μ-harmonic functions, for all spherical measures μ, coincide with the space H of L.U.C. harmonic functions. The Poisson pairs for spherical measures are all isomorphic.

Proof: For a fixed $g \in \mathbb{G}$ let

$$F(g') = \int_{\mathbb{K}} f(gkg')\,dk.$$

Then

$$\int_{\mathbb{G}} F(g'g'')\,d\mu(g'') = \int_{\mathbb{G}} \int_{\mathbb{K}} f(gkg'g'')\,dk\,d\mu(g'') =$$

$$\int_{\mathbb{K}} \int_{\mathbb{G}} f(gkg'g'')\,d\mu(g'')\,dk = \int_{\mathbb{K}} f(gkg')\,dk = F(g').$$

Thus $F(g')$ is μ-harmonic and by its definition $F(kg') = F(g'')$ for every $g \in \mathbb{G}$ and $k \in \mathbb{K}$. Hence by lemma 3.2. $F(g')$ is a constant. Taking $g' = e$ and recalling that $f(gk) = f(g)$ we have $F(e) = f(g)$. It follows that for every $g,\ g' \in \mathbb{G}$,

$$f(g) = \int_{\mathbb{K}} f(gkg')\,dk\,;$$

by theorem 3.1.(3) f is harmonic. //

We remark that another characterization of harmonic functions is given in [13]. By theorem 3.1.(1), for a harmonic function f,

$$\tilde{f}(g) = \int_X \phi(gx)\, dm(x) = \int_X \phi(x) \frac{dgm}{dm}(x)\, dm(x) =$$

$$= \int_X \phi(x)\, P(g,x)\, dm(x),$$

where $P(g,x)$ is the Radon-Nikodym derivative of gm with respect to m. (This is the analog of the classical Poisson kernel.) Since $km = m$, $P(g,x)$ is defined on $\mathbb{G}/\mathbb{K} \times X$ and it is C^∞ function in its $\mathbb{G}/\mathbb{K} = D$ variable. This implies that f is C^∞ on D. A <u>Laplace operator</u> on D is an elliptic second order differential operator on D which vanishes for constants and is $\mathbb{G}$-invariant. The following theorem is proved in [14, page 366].

<u>3.4. THEOREM</u>: <u>If $f(z)$ is a twice differentiable bounded function on D satisfying $\Delta f = 0$ for some Laplace operator Δ, then $f(z)$ is harmonic. Conversely if $f(z)$ is harmonic on D then it is C^∞ on D and $\Delta f = 0$ for any Laplace operator Δ.</u>

VI.4. THE POISSON SPACE CORRESPONDING TO $\mathbb{G}$ AND A SPHERICAL MEASURE.

Let μ be an absolutely continuous measure on $\mathbb{G}$ and let (B,ν) be the corresponding Poisson pair.

<u>4.1. LEMMA</u>: <u>B is a homogeneous space of $\mathbb{G}$ and $\mathbb{K}$ is already transitive on B. Moreover B can be represented as $\mathbb{G}/\mathbb{H}$ where $\mathbb{H} \subset \mathbb{P} = \mathbb{M}\mathbb{A}\mathbb{N}$.</u>

<u>Proof</u>: If $\mathbb{K}$ is not transitive on B then there exists two disjoint $\mathbb{K}$-invariant closed subsets B_1 and B_o of B. Let ϕ be a continuous function on B such that $\phi(x) = 1$ for all $x \in B_1$ and $\phi(x) = 0$ whenever $x \in B_o$. Let

$$\psi(x) = \int_{\mathbb{K}} \phi(kx)\, dk ,$$

then ψ is a continuous $\mathbb{K}$-invariant and non-constant function on B.

Put

$$f(g) = \int_B \psi(gx)\, d\nu(x) ,$$

then f is μ-harmonic and $f(kg) = f(g)$ for every $k \in \mathbb{K}$. Hence by lemma 3.2. $f(g)$ is a constant which is a contradiction. Thus $\mathbb{K}$ is transitive on B. By corollary V.4.5. $\pi_S(\mathbb{G}) = X = \mathbb{G}/\mathbb{P}$ is an equivariant image of $B = \mathbb{G}/\mathbb{H}$ hence we can assume that $\mathbb{H} \subset \mathbb{P} = \mathbb{M}\mathbb{A}\mathbb{N}$. //

Using this lemma and the fact that ν is contractible one can show that <u>for an absolutely continuous probability measure</u> μ, B <u>is a finite covering space of</u> X; <u>i.e.</u>, $\mathbb{H}$ <u>has a finite index in</u> $\mathbb{P}$. <u>If moreover for some</u> n <u>the support of</u> $\mu^n = \mu * \cdots * \mu$ <u>contains a neighborhood of</u> e <u>in</u> $\mathbb{G}$ <u>then</u> B = X. (Theorem 5.2 and 5.3 of [13]).

By theorem 3.3. the Poisson space (B,ν) associated with a spherical measure μ does not depend on μ. Thus if we choose spherical μ whose support equals $\mathbb{G}$, then we can deduce that (B,ν) is isomorphic to (X,m). Thus <u>for every spherical</u> μ <u>the corresponding Poisson pair is</u> $(\pi_S(\mathbb{G}),m)$, the maximal boundary of $\mathbb{G}$ with its unique $\mathbb{K}$-invariant measure. In particular the Banach space H of L.U.C. harmonic functions is isometrically isomorphic to $C(X)$.

We call a function h on D an H-<u>function</u> if it is the restriction to D of an affine continuous function on M. (Every continuous affine function on M is given by $f(\nu) = \int_X \phi\, d\nu$ where $\phi = f|X$.)

4.2. <u>THEOREM</u>: <u>A function</u> h <u>on</u> D <u>is an</u> H-<u>function iff</u> $h \in H$; <u>i.e., the</u> L.U.C. <u>harmonic functions are the</u> H-<u>functions</u>.

<u>Proof</u>: Let h be an H-function, let f be a continuous affine function on M such that $f|D = h$ and let $\phi = f|X$. Now

$$\tilde{h}(g) = h(gm) = f(gm) = \int_X \phi(x)\, dgm = (L_m\phi)(g).$$

Thus $\tilde{h} = L_m\phi$ and $\tilde{h}$ is L.U.C. μ-harmonic function for every probability measure μ on $\mathbb{G}$ which satisfies $\mu * m = m$. (V.1.2.) In particular $\tilde{h}$ is μ-harmonic for every spherical μ.

Conversely let h be L.U.C. harmonic function. Then $\tilde{h}$ is μ-harmonic for some spherical μ hence by the property of the Poisson space (B,ν) associated with $\mathbb{G}$, μ there exists a continuous function ϕ on B such that $\tilde{h} = L_\nu\phi$. But for spherical μ $(B,\nu) = (X,m)$, thus $\tilde{h} = L_m\phi$ for some continuous function ϕ on X, and if we define

$$f(\lambda) = \int_X \phi \, d\lambda \qquad (\lambda \in M),$$

then f is an affine continuous function on M and $f|D = h$. //

In [5] Azencott proves several generalizations of these results for example he shows that two etalée probability measures on $\mathbb{G}$ whose supports generate the same closed sub-semigroup of $\mathbb{G}$ have isomorphic Poisson spaces. (For the definition of an etalée measure see section V.5. above).

VI.5. THE CASE $\mathbb{G} = \mathbb{SL}(2,\mathbb{R})$.

We have seen already that for $\mathbb{G} = \mathbb{SL}(2,\mathbb{R})$, $\mathbb{G} = \mathbb{K}\mathbb{S}$ where

$$\mathbb{K} = \left\{ \begin{pmatrix} \cos\theta & -\sin\theta \\ \sin\theta & \cos\theta \end{pmatrix} \;\middle|\; 0 \le \theta \le 2\pi \right\} \qquad \text{and}$$

$$\mathbb{S} = \left\{ \begin{pmatrix} a & b \\ 0 & a^{-1} \end{pmatrix} \;\middle|\; a > 0 \right\},$$

is an Iwasawa decomposition and the corresponding $\mathbb{P}$ is

$$\mathbb{P} = \left\{ \begin{pmatrix} a & b \\ 0 & a^{-1} \end{pmatrix} \;\middle|\; a \ne 0 \right\}.$$

Let $\mathbb{G}$ act on $\mathbb{C}$ by linear fractional transformations, i.e.

$$\begin{pmatrix} a & b \\ c & d \end{pmatrix} z = \frac{az + b}{cz + d} .$$

This action of $\mathbb{G}$ is transitive on

$$U = \{z \mid \text{Im } z > 0\} \quad \text{and also on}$$

$$L = \{z \mid \text{Im } z = 0\} \cup \{\infty\} .$$

Moreover $\{g \mid gi = i\} = \mathbb{K}$ and $\{g \mid g\infty = \infty\} = \mathbb{P}$, and thus U is isomorphic to D and L is isomorphic to X. So we have canonical homeomorphisms of U and L in M and we will show that they define a homeomorphism of $\{z \mid \text{Im } z \geq 0\} \cup \{\infty\}$ into $D \cup X$ as a subset of M. Because of the transitivity of $\mathbb{G}$ on L it suffices to show that $g_n i \to \infty$ iff $g_n m \to \delta_\infty$ where $g_n = \begin{pmatrix} a_n & b_n \\ 0 & a_n^{-1} \end{pmatrix}$ and m is the unique $\mathbb{K}$-invariant measure on L. By the bounded convergence theorem it suffices to show that $g_n z \to \infty$ for almost all $z \in L$. If $|a_n| \leq M$ for some M and all n, then $|g_n i|^2 = a_n^4 + a_n^2 b_n^2 \to \infty$ implies $a_n^2 b_n^2 \to \infty$ and for $z \in \mathbb{R}$

$$|g_n z| \geq |a_n b_n| - |a_n^2 z| \to \infty .$$

If $|a_n| \to \infty$ and $b_n/a_n \to \beta$ then for $z \neq -\beta$, $|g_n z| = |a_n|^2 |z + b_n/a_n| \to \infty$. Hence in these special cases $g_n m \to \delta_\infty$. Since M is compact the general case follows because every subsequence has a subsequence of one of these types. Finally, since $L \cup U$ is compact this one to one continuous map is a homeomorphism.

Next we shift to the unit disc. Let $A = \begin{pmatrix} 1 & -i \\ 1 & i \end{pmatrix}$ and let $\tilde{U} = \{z \mid |z| < 1\}$, $\tilde{L} = \{z \mid |z| = 1\}$ and $\tilde{\mathbb{G}} = \left\{ \begin{pmatrix} \alpha & \bar{\beta} \\ \beta & \bar{\alpha} \end{pmatrix} \middle| \ \alpha\bar{\alpha} - \beta\bar{\beta} = 1 \right\}$.
Then $\tilde{U} = AU$, $\tilde{L} = AL$, $\tilde{\mathbb{G}} = A\mathbb{G}A^{-1}$ and $Ai = 0$. Let $\tilde{\mathbb{K}} = A\mathbb{K}A^{-1}$ then the unique $\tilde{\mathbb{K}}$ invariant measure $\tilde{m}$ on $\tilde{L}$ is the Lebesgue measure on the unite circle. If we put $z = \tilde{g}0$ for $\tilde{g} \in \tilde{\mathbb{G}}$ then for

a harmonic function on $D \simeq U \simeq \tilde{U}$ we have

$$f(z) = f(\tilde{g}0) = \int_{\tilde{L}} \phi(\tilde{g}\zeta)\, d\zeta = \int_{\tilde{L}} \phi(\zeta) \left| \frac{d\tilde{g}^{-1}}{d\zeta} \right| d\zeta$$

where $\phi \in L^{\infty}(\tilde{L},\tilde{m})$ and $d\zeta = d\tilde{m}$. Letting $\tilde{g} = \begin{pmatrix} \alpha & \overline{\beta} \\ \beta & \overline{\alpha} \end{pmatrix}$ we have $\tilde{g}^{-1} = \begin{pmatrix} \overline{\alpha} & -\overline{\beta} \\ -\beta & \alpha \end{pmatrix}$ and

$$\left| \frac{d\tilde{g}^{-1}}{d\zeta} \right| = \frac{1}{|-\beta\zeta+\alpha|^2} = \frac{1}{|-\beta+\overline{\zeta}\alpha|^2} = \frac{1}{|\overline{\alpha}\zeta-\overline{\beta}|^2}$$

for $|\zeta| = 1$. The classical Poisson kernel is

$$\frac{1-z\overline{z}}{|\zeta-z|^2} .$$

This simplifies to $\dfrac{1}{|\overline{\alpha}\zeta-\overline{\beta}|^2}$ when we substitute $z = \overline{\beta}/\overline{\alpha} = \tilde{g}0$.

Finally we have the classical Poisson formula

$$f(z) = \int_{|\zeta|=1} \phi(\zeta)\, \frac{1-z\overline{z}}{|\zeta-z|^2}\, d\zeta.$$

CHAPTER VII

EXPONENTIAL GROWTH, AMENABILITY AND FREE SUBGROUPS ON TWO GENERATORS

Let T be a discrete group; the following open problem are well known.

A. Is it true that T is non-amenable iff it contains a free subgroup on two generators?

B. If T is finitely generated and not of exponential growth (see section one for the definition), is T necessarily a finite extension of a nilpotent group?

J. Tits has shown [40] that if T is finitely generated and has a faithful linear complex representation, then either T contains a free subgroup or it is a finite extension of a solvable group. So that in the category of finitely generated linear groups the answer to A is affirmative.

J. Milnor and J. Wolf have shown [32],[44], that the answer to probelm B is affirmative when T is a solvable group. Thus by the Tits' result the answer is also affirmative for finitely generated linear groups.

Corollary 3.2. below, states that a finitely generated T which is not of exponential growth is amenable. Thus this corollary supports both conjectures A and B. In section one of this chapter we define exponential growth for a finitely generated group and show that the definition does not depend on the set of generators. In section two we prove a characterization of strongly proximal flows which is used in section 3 to prove corollary 3.2. Using similar methods we show that any two non-commuting homeomorphisms of the circle of which at least one acts minimally on the circle, generates a subgroup of homeomorphism which is of exponential growth, and that for many groups one can deduce the existence of a free subgroup by considering certain non-trivial minimal flows which they admit.

We remark that the notion of exponential growth has been generalized to locally compact topological goups and interesting characterizations have been given of connected Lie groups not of exponential growth. ([28] and [24]). In [24] Y. Guivarch gives a short proof, due to A. Avez of the following statement: If T is a locally compact topological group not of exponential growth then T is amenable. This is of course more general than corollary 3.2. yet we think that our proof of this corollary is interesting enough to present here. The main sources for this chapter are [32] and [20].

VII.1. EXPONENTIAL GROWTH

Let T be a finitely generated group and let $A = \{a_1, \cdots, a_n\}$ be some set of generators of T. For each $t \in T$ let $\ell_A(t)$ be the minimum number of elements of $A \cup A^{-1}$ required to express t, and for each positive integer k let $\gamma_A(k)$ be the number of elements in T such that $\ell_A(t) \leq k$. (By convention $\ell_A(e) = 0$ and $\gamma_A(0) = 1$.) The group T has <u>exponential growth</u> if for some finite set A of generators and some $a > 1$

$$\gamma_A(k) \geq a^k \qquad \text{for all } k.$$

<u>1.1. LEMMA</u>: <u>The</u> $\lim\limits_{k\to\infty} \gamma_A(k)^{1/k}$ <u>exists and equals</u> $\inf\limits_k \{\gamma_A(k)^{1/k}\}$.

<u>Proof</u>: We write γ for γ_A. Note that

$$\gamma(p+q) \leq \gamma(p) \cdot \gamma(q).$$

Fix a positive integer m and let $h = \left[\frac{k}{m}\right] + 1$, where $[\cdot]$ denotes integral part.

Then we have

$$\gamma(k) \le \gamma(hm) \le \gamma(m)^h \le \gamma(m)^{k/m}\gamma(m)$$

and

$$\gamma(k)^{1/k} \le \gamma(m)^{1/m}\gamma(m)^{1/k}.$$

Consequently, for all m

$$\overline{\lim_{k\to\infty}}\ \gamma(k)^{1/k} \le \gamma(m)^{1/m}$$

from which it follows that

$$\overline{\lim_{k\to\infty}}\ \gamma(k)^{1/k} \le \inf_m \gamma(m)^{1/m} \le \underline{\lim}\ \gamma(k)^{1/k}. //$$

We denote this limit by c_A.

1.2. LEMMA: *If A and B are finite sets of generators of T, then there exists a positive integer d such that $c_A \le c_B^d$.*

Proof: Let $d = \max\{\ell_B(t) \mid t \in A\}$. Then

$$\gamma_A(k) \le \gamma_B(dk) \le (\gamma_B(k))^d,$$

which implies that $c_A \le c_B^d$. //

The following proposition is an immediate consequence of these two lemmas.

1.3. PROPOSITION: *Let A and B be finite sets of generators of T. Then $c_A > 1$ iff for some $a > 1$, $\gamma_A(k) \ge a^k$ for all k. There exists such an a iff there exists $b > 1$ such that $\gamma_B(k) \ge b^k$ for all k.*

EXAMPLES: 1. Let $T = \mathbf{Z} \oplus \mathbf{Z}$ and let $A = \{(1,0),(0,1)\}$. It is easy to check that $\gamma_A(k) - \gamma_A(k-1) = 4k$ and thus

$$\gamma_A(k) = 1 + \sum_{m=1}^{k} \gamma_A(m) - \gamma_A(m-1) = 1 + \sum_{m=1}^{k} 4m$$

$$= 2k^2 + 2k + 1.$$

Therefore $\mathbb{Z} \oplus \mathbb{Z}$ does not have an exponential growth.

2. Let F_2 be the free group on two generators a and b. Let $A = \{a,b\}$. Consider elements of the form $a^{\varepsilon_1} b^{\varepsilon_2} a^{\varepsilon_3} \cdots c^{\varepsilon_k}$ where $\varepsilon_i = \pm 1$ and c is either a or b according as k is odd or even. Now we have $\gamma(k) \geq 2^k$ and F_2 has exponential growth.

3. Let S be the group of 2×2 matrices generated by the matrices

$$x = \begin{pmatrix} \sqrt{2} & 0 \\ 0 & \sqrt{2}^{-1} \end{pmatrix} \quad \text{and} \quad y = \begin{pmatrix} 1 & 1 \\ 0 & 1 \end{pmatrix}.$$

Clearly $[S,S]$, the commutator group of S, consists of matrices of the form $\begin{pmatrix} 1 & a \\ 0 & 1 \end{pmatrix}$, hence S <u>is solvable of degree two</u>. Next we show that for a given n all the words

$$(*) \qquad xy^{\varepsilon_1} xy^{\varepsilon_2} \cdots xy^{\varepsilon_n}$$

where $\varepsilon_i = 0,1$, represent distinct elements of S. This will imply of course that S <u>has an exponential growth</u>. Notice that $xyx^{-1} = y^2$. (As a matter of fact one can show that S is isomorphic to the free group on two generators modulo this relation.)

First we prove by induction on n that in each element $\begin{pmatrix} a & b \\ 0 & a^{-1} \end{pmatrix}$ of S which has the representation $(*)$, $b < 2a$. Indeed for $n = 1$ the matrices of the form $(*)$ are

$$\begin{pmatrix} \sqrt{2} & 0 \\ 0 & \sqrt{2}^{-1} \end{pmatrix} \quad \text{and} \quad \begin{pmatrix} \sqrt{2} & \sqrt{2} \\ 0 & \sqrt{2}^{-1} \end{pmatrix}$$

and $\sqrt{2} < 2\sqrt{2}$.

If $\begin{pmatrix} a_1 & b_1 \\ 0 & a_1^{-1} \end{pmatrix}$ has a representation $(*)$ of length $n+1$ then

then there exists a matrix $\begin{pmatrix} a & b \\ 0 & a^{-1} \end{pmatrix}$ of the form (*) of length n such that

$$\begin{pmatrix} a_1 & b_1 \\ 0 & a_1^{-1} \end{pmatrix} = \begin{pmatrix} a & b \\ 0 & a^{-1} \end{pmatrix}\begin{pmatrix} \sqrt{2} & 0 \\ 0 & (\sqrt{2})^{-1} \end{pmatrix} = \begin{pmatrix} a\sqrt{2} & \frac{b\sqrt{2}}{2} \\ 0 & (a\sqrt{2})^{-1} \end{pmatrix}$$

or

$$\begin{pmatrix} a_1 & b_1 \\ 0 & a_1^{-1} \end{pmatrix} = \begin{pmatrix} a & b \\ 0 & a^{-1} \end{pmatrix}\begin{pmatrix} \sqrt{2} & \sqrt{2} \\ 0 & (\sqrt{2})^{-1} \end{pmatrix} = \begin{pmatrix} a\sqrt{2} & a\sqrt{2} + \frac{b\sqrt{2}}{2} \\ 0 & (a\sqrt{2})^{-1} \end{pmatrix}.$$

We assume $b < 2a$ and then in the first case

$$b' = \frac{b\sqrt{2}}{2} < \frac{2a\sqrt{2}}{2} = a\sqrt{2} = a',$$

and in the second

$$b' = \sqrt{2}\left(a + \frac{b}{2}\right) < \sqrt{2}\left(a + \frac{2a}{2}\right) = 2a\sqrt{2} = 2a'.$$

Next assume that all the $2n$ words of the form (*) represent different elements of S and suppose that

$$(\#) \qquad xy^{\varepsilon_1}xy^{\varepsilon_2}\cdots xy^{\varepsilon_{n+1}} = xy^{\theta_1}xy^{\theta_2}\cdots xy^{\theta_{n+1}}.$$

If $\varepsilon_{n+1} = \theta_{n+1}$ we have

$$xy^{\varepsilon_1} \cdots xy^{\varepsilon_n} = xy^{\theta_1} \cdots xy^{\theta_n}$$

and by our assumption also $\varepsilon_i = \theta_i$ $1 \le i \le n$. Let therefore $\varepsilon_{n+1} = 1$ and $\theta_{n+1} = 0$. Multiplying (#) by x^{-1} on the right we have,

$$xy^{\varepsilon_1} \cdots xy^{\varepsilon_n}y^2 = xy^{\theta_1} \cdots xy^{\theta_n}.$$

Let

$$xy^{\varepsilon_1} \cdots xy^{\varepsilon_n} = \begin{pmatrix} a & b \\ 0 & a^{-1} \end{pmatrix}$$

and $xy^{\theta_1}\cdots xy^{\theta_n} = \begin{pmatrix} a & c \\ 0 & a^{-1} \end{pmatrix}$; then

$$\begin{pmatrix} a & c \\ 0 & a^{-1} \end{pmatrix} = \begin{pmatrix} a & b \\ 0 & a^{-1} \end{pmatrix}\begin{pmatrix} 1 & 2 \\ 0 & 1 \end{pmatrix} = \begin{pmatrix} a & 2a+b \\ 0 & a^{-1} \end{pmatrix}.$$

Thus $c = 2a+b$ which is a contradiction to the inequality $c < 2a$ since b is non-negative.

VII.2 A CHARACTERIZATION OF STRONG PROXIMALITY WHICH DOES NOT DEPEND UPON MEASURES.

Let $\{x_i\}_{i=1}^n$ be a finite sequence of points. We say that this sequence has property P ε-almost everywhere if at most $[n\varepsilon]$ terms of the sequence do not have it, where $[\cdot]$ denotes the integral part of a real number. We use this definition to obtain the following characterization of strongly proximal minimal flows.

2.1 PROPOSITION: *Let (T,X) be a minimal flow and let $x_o \in X$. Then (T,X) is strongly proximal if and only if for every $\varepsilon > 0$ and every neighbourhood V of x_o there exists a finite subset $F \subset T$ such that given any finite sequence $\{x_i\}_{i=1}^n$ of points in X there exists t in F for which ε-almost every tx_i is in V.*

Proof: Suppose that X is strongly proximal. Let ψ be a continuous function on X such that $0 \le \psi(x) \le 1$, $\psi(x_o) = 1$ and $\psi(x) = 0$ if $x \notin V$, and let $U = \{u \in M(X) \mid |u(\psi) - \delta_{x_o}(\psi)| < \varepsilon\}$. Because $(T,M(X))$ is proximal and δ_{x_o} belongs to the unique minimal set in $M(X)$, given $u \in M(X)$ there exists a $t_u \in T$ such that $t_u u \in U$. Let V_u be an open neighborhood of u in $M(X)$ such that $t_u V_u \subset U$ and choose a finite subcovering $\{V_{u_1},\cdots,V_{u_k}\}$ of $M(X)$. If we denote $t_i = t_{u_i}$ $(i = 1,\cdots,k)$ and $F = \{t_i\}_{i=1}^k$, then for every $u \in M(X)$ there exists $t \in F$ such that $tu \in U$.

Let $\{x_i\}_{i=1}^n$ be a finite subset of X and let $u = \frac{1}{n}\sum_{i=1}^n \delta_{x_i}$. We shall complete the proof by showing that if $tu \in U$ then ε-almost every tx_i is in V. Let a be the number of terms in $\{tx_i\}_{i=1}^n$ not in V. On the one hand

$$1 - \frac{1}{n}\sum_{i=1}^n \psi(tx_i) \geq 1 - \frac{n-a}{n} = \frac{a}{n}.$$

On the other hand,

$$1 - \frac{1}{n}\sum_{i=1}^n \psi(tx_i) = |\delta_{x_o}(\psi) - tu(\psi)| < \varepsilon.$$

Thus $a/n < \varepsilon$ or $a \lesssim [n\varepsilon]$.

Now suppose the condition holds. Let M_o be the set of all rational convex combinations of point masses; let $\{\psi_1,\cdots,\psi_q\}$ be a finite set of continuous functions on X and let $\varepsilon > 0$ be given. Set $\varepsilon' = \varepsilon/3M$ where $M = \max\{\|\psi_i\| \mid 1 \leq i \leq q\}$ and set $V = \{x \in X \mid |\psi_i(x) - \psi_i(x_o)| < \varepsilon'$ for $1 \leq i \leq q\}$. Let F be a finite subset of T given by the hypothesis for V and ε'. Given $u \in M_o$ we can assume that $u = \frac{1}{n}\sum \delta_{x_i}$ by allowing repeats. There exists $t \in F$ such that ε'-almost every tx_i is in V. Without loss of generality we can assume that $tx_i \notin V$ iff $1 \leq i \leq p \leq [n\varepsilon']$. Hence for $1 \leq k \leq q$

$$|tu(\psi_k) - \delta_{x_o}(\psi_k)| = |\frac{1}{n}\sum_{i=1}^n \psi_k(tx_i) - \psi_k(x_o)| \leq$$

$$\frac{2pM}{n} + \frac{n-p}{n}\varepsilon' \leq \frac{2[n\varepsilon']M}{n} + \varepsilon' \leq \frac{2}{3}\varepsilon + \frac{1}{3}\varepsilon = \varepsilon.$$

There exists an index (an element of the uniform structure) β on $M(X)$ such that $(\lambda,\nu) \in \beta$ implies $(t\lambda,t\nu) \in \alpha$, where $\alpha = \{(\lambda,\nu) \mid |u(\psi_k) - \nu(\psi_k)| < \varepsilon,\ 1 \leq k \leq q\}$, for all $t \in F$. Given $\nu \in M(X)$ there exists $u \in M_o$ such that $(u,\nu) \in \beta$. It follows that there exists $t \in F$ satisfying $|t\nu(\psi_k) - \delta_{x_o}(\psi_k)| < 2\varepsilon$ for $1 \leq k \leq q$ and there exists a net $\{t_i\}$ in T such that $\lim t_i\nu = \delta_{x_o}$ which is equivalent to (T,X) being strongly proximal.//

VII.3 A NON-AMENABLE GROUP CONTAINS A FINITELY GENERATED SUBGROUP OF EXPONENTIAL GROWTH.

Using the characterization of strongly proximal minimal flows given in section 2 and theorem III.3.1. we now prove the following theorem.

3.1. THEOREM: Let T be a non-amenable group, then T contains a finitely generated subgroup of exponential growth.

Proof: By Theorem III.3.1. we know that there exists a non-trivial minimal strongly proximal flow (T,X). Let $x_o \in X$, $t \in T$ such that $tx_o \neq x_o$, V an open neighborhood of x_o such that $V \cap tV = \emptyset$ and $0 < \varepsilon < 1/4$. Now let F be a finite subset of T which corresponds to ε and V as in proposition 2.1. and let S be the subgroup of T generated by $A = F \cup \{t\}$. If $w_1,\cdots,w_q$ are distinct elements of S, then there exists $s \in F$ such that ε-almost every sw_ix_o is in V. It follows that $\{sw_1,\cdots,sw_q\} \cup \{tsw_1,\cdots,tsw_q\}$ contains at least $2(q-[q\varepsilon])$ distinct elements because $V \cap tV = \emptyset$. If the w_i's are the elements of S such that $\ell_A(w_i) \leq k$, then

$$\gamma_A(k+2) \geq 2(\gamma_A(k) - [\gamma_A(k)\varepsilon]) \geq$$

$$2\gamma_A(k) - 2[\gamma_A(k)/4] \geq 2\gamma_A(k) - \tfrac{1}{2}\gamma_A(k) = \tfrac{3}{2}\gamma_A(k),$$

and S has exponential growth.//

3.2. COROLLARY: If T is finitely generated group which is not of exponential growth then T is amenable.

Proof: If a finitely generated group has a subgroup of exponential growth, then it is of exponential growth.//

If (T,X) is a minimal flow and U is an open subset of X then there exists a finite set $F \subset T$ such that $FU = X$. Let $n(U)$ be the smallest cardinality of all such F's. Let $m(U)$ be the largest cardinality of a subset D of T such that $s,t \in D$ implies $sU \cap tU = \emptyset$.

3.3. PROPOSITION: Let (T,X) be a minimal flow. If there exists an open subset U of X such that $m(U) > n(U)$ then T contains a finitely generated subgroup of exponential growth.

Proof: Let F be a subset of T of cardinality $n(U)$ such that $FU = X$ and let D be a finite subset of T of cardinality m greater than $n(U)$, such that $s,t \in D$ implies $sU \cap tU = \emptyset$. Let S be the subgroup of T generated by $A = F \cup D$. If $w_1,\cdots,w_q$ are distinct elements of S and $n(U)\nmid q$, then at least $[q/n(U)] + 1$ of the points $w_1x_o,\cdots,w_qx_o$ are in tU for some t in F. Therefore $\{st^{-1}w_i \mid s \in D, 1 \le i \le q\}$ contains at least $m([q/n(U)] + 1)$ distinct elements. Now as in the proof of theorem 3.1.

$$\gamma_A(k+2) \ge m([\gamma_A(k)/n(U)] + 1) \ge \frac{m}{n(U)}\, \gamma_A(k)$$

and S is of exponential growth. Similarly, when $n(U)|q$ $\gamma_A(k+2) \ge \frac{m}{n(U)}\, \gamma_A(k)$. //

3.4. EXAMPLE: Let X be the unit circle $\{t \in C \mid |z| = 1\}$. Let ϕ and ψ be a pair of non-commuting homeomorphisms of X onto itself. Assume that (ϕ,X) is minimal and that ψ is orientation preserving then the group T of homeomorphisms of X generated by ϕ and ψ has an exponential growth.

To prove this we first observe that by [31] we can assume that is an irrational rotation of the circle. Since ψ does not commute with ϕ, there exists a closed connected arc C such that $\psi(C)$ is shorter than C. Using iterates of ϕ we can find $\theta \in T$ $(\theta = \phi^n\psi$

for some n) such that $\theta(C) \subset C$ and $\theta(C)$ does not contain the end points of C. From this it is easy to deduce that there exists an open arc U for which the lengths of the arcs $\theta^k(U)$ tends to zero when k tends to infinity. It is clear now that $n(U) < m(U) \leq \infty$, so that by proposition 3.3. T has an exponential growth.

An extremely proximal flow is a minimal flow X which contains more than two points such that whenever A is a closed proper subset of X and V is an open non-empty subset of X, there exists $t \in T$ for which $tA \subset V$.

3.5. PROPOSITION: *If T admits a non-trivial extremely proximal flow, then T contains a free subgroup on two generators.*

Proof: Let U and V be disjoint non-empty open subsets in X, where X is a non-trivial extremely proximal flow of T, such that $U \cup V \neq X$. Let U_1, U_2 and V_1, V_2 be disjoint non-empty subsets in U and V respectively. Then there exists s and t such that $t(X\setminus U_1) \subset U_2$ and $s(X\setminus V_1) \subset V_2$. Hence we also have $t^{-1}(X\setminus U_2) \subset U_1$ and $s^{-1}(X\setminus V_2) \subset V_1$. Let S be the subgroup of T generated by s and t and let $x \in X\setminus (U \cup V)$. If w is a reduced word in t and s then it is easy to see that $wx \in U_1 \cup U_2 \cup V_1 \cup V_2$. Thus $wx \neq x$ and $w \neq e$; i.e. S is free.//

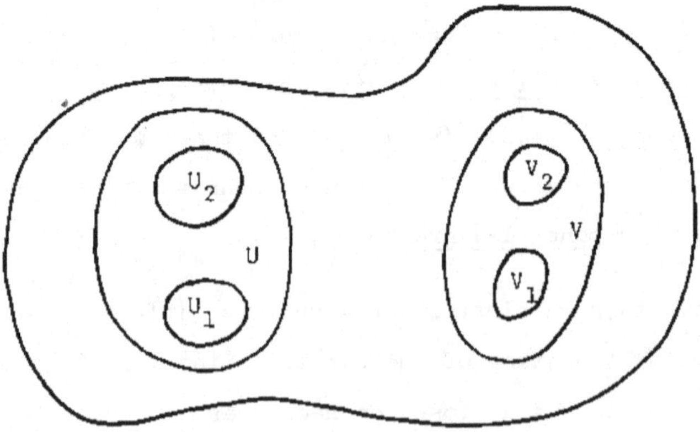

3.6. EXAMPLE: Let T be a Fuchsian group of the first kind and let X be the unit circle then (T,X) is extremely proximal because the end points of axis of hyperbolic transformations in T are dense in X. Thus every such a group contains a free subgroup.

CHAPTER VIII

THE GENERALIZED BOHR COMPACTIFICATION. I

We say that a continuous homomorphism of a topological group T into a compact topological group K, $T \xrightarrow{\psi} K$, is a compactification of T if $\psi(T)$ is dense in K. If $T \xrightarrow{\psi} K$ is a compactification such that whenever $T \xrightarrow{\phi} L$ is another compactification there exists a homomorphism $\theta: K \to L$ such that $\phi = \theta \circ \psi$, then $T \xrightarrow{\psi} K$ is called the universal compactification of T or the Bohr compactification of T. It is easy to see that the Bohr compactification exists and it is unique up to an isomorphism.

A compactification of T, say, $T \xrightarrow{\psi} L$ gives rise to an equicontinuous flow (T,L) defined by $t\ell = \psi(t)\ell$ $(t \in T, \ \ell \in L)$, and it follows from theorem I.3.3.(5). that the Bohr compactification of T is the universal minimal equicontinuous flow. We denote this flow by $(T,\Gamma(T))$.

If T is abelian, $\Gamma(T)$ is a very usefull object in harmonic analysis on T as well as in topological dynamics. However for non-abelian groups $\Gamma(T)$ may be trivial even when T is quite big. For example it follows from a theorem of C. C. Moore [33,34] that if T is a connected locally compact group whose action on $\Gamma(T)$ is effective then T is a compact extension of a solvable group. In particular for a connected semi-simple Lie group $\mathfrak{G}$ without compact components $\Gamma(\mathfrak{G})$ is trivial.

We would like to find a substitution $\Sigma(T)$ for $\Gamma(T)$ which will reduce to $\Gamma(T)$ for abelian T and will be non-trivial for a much larger class of topological groups than for which $\Gamma(T)$ is non-trivial. In particular we would like $\Sigma(\mathfrak{G})$ to be non-trivial for a connected semi-simple Lie group $\mathfrak{G}$ without compact components.

We have seen (Chapter I, lemma 3.3.) that if x and y are proximal points in a minimal flow X and $x \neq y$, then there is no automorphisms of X which maps x onto y. Thus a minimal flow in which for every pair of points x and y there exists an automorphism ϕ of X such that $\phi(x)$ is proximal to y, has as many automorphisms as one can hope for. We say that such a flow is a regular flow. An equivalent characterization of a regular flow is that it is isomorphic to a minimal ideal in its enveloping semigroup. [2].

Now we call a regular flow (T,X) a compactification flow of T if the group of all automorphisms of (T,X) (say with the topology of point-wise convergence) is a compact Hausdorff topological group. In this case we say that this compact group is a generalized compactification of T. It turns out that a flow (T,X) is a compactification flow iff it is a group extension (see section 1) of a proximal flow (proposition 2.1.). We now say that a minimal flow is a strong compactification flow if it is a group extension of a strongly proximal flow and we call the compact group of automorphisms of this flow, a generalized strong compactification of T.

Finally we let $\Sigma(T)$, $\{\Sigma_S(T)\}$ be the compact group of automorphisms of the universal {strong} compactification flow of T, and we call it the generalized {strong} Bohr compactification.

In this chapter we identify (using ideas which appear in [15]) $\Sigma_S(\mathbb{G})$ for a connected semi-simple Lie group with a finite center. Of course $\Sigma_S(\mathbb{G}) = \Sigma(\mathbb{G})$ if $\pi_S(\mathbb{G}) = \pi(\mathbb{G})$ (see IV Section 6). In any case it is clear that $\Sigma(\mathbb{G})$ is non-trivial since $\Sigma_S(\mathbb{G})$ is so. In the next chapter we shall identify $\Sigma(T)$, for a discrete group T, in terms of βT. In particular we shall show there, that $\Sigma(T) = \Gamma(T)$ whenever $\pi(T)$ is trivial.

VIII.1 GROUP EXTENSIONS AND ALMOST PERIODIC EXTENSIONS

Let (T,Z) be a minimal flow and let $\chi: (T,Z) \to (T,Y)$ be a homomorphism. We say that χ is a group extension if there exists a compact Hausdorff topological group K such that

(i) K acts jointly continuously on Z (on the right)

(ii) for every $z \in Z$, $\chi^{-1}(\chi(z)) = zK$

(iii) for all z, k, and t $(tz)k = t(zk)$.

If $zk = z$ for some z and k then $zk = z$ for all $z \in Z$, hence we can assume that K acts freely (i.e., $zk = z$ for some z implies $k = e$). We denote the group extension by $(T,Z,K) \xrightarrow{\chi} (T,Y)$.

Let $(T,X,x_o) \xrightarrow{\phi} (T,Y,y_o)$ be a homomorphism of pointed minimal flows. We say that ϕ is an *almost periodic* homomorphism (or extension) if there exist a group extension $(T,Z,z_o,K) \xrightarrow{\chi} (T,Y,y_o)$ and a homomorphism $(T,Z,z_o) \xrightarrow{\psi} (T,X,x_o)$ so that the following diagram is commutative:

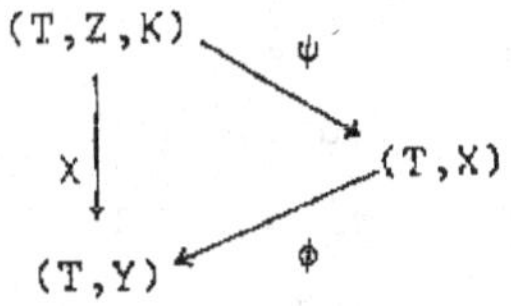

Clearly every group extension and hence also every almost periodic extension is distal. (Since every $k \in K$ can be considered as an automorphism of (T,Z) this follows for example from II.3.3.).

1.1. LEMMA: *In the above diagram let* $K_o = \{k \in K \mid \psi(z_o k) = x_o\}$, *then* $(T,Z,K_o) \xrightarrow{\psi} (T,X)$ *is a group extension. If moreover* $(T,X,L) \to (T,Y)$ *is also a group extension then there exists a homomorphism* $\theta: K \to L$ *onto* L *such that* $\ker \theta = K_o$ *and such that* $\psi(zk) = \psi(z)\theta(k)$ *for every* $z \in Z$ *and* $k \in K$.

Proof: Let $k, k' \in K_o$ and let $\{t_i\}$ be a net in T such that $\lim t_i z_o = z_o k$ then

$$\psi(z_o kk') = \psi(\lim(t_i z_o k')) = \lim t_i \psi(z_o k') =$$

$$= \lim t_i x_o = \lim \psi(t_i z_o) = \psi(z_o k) = x_o .$$

Thus $kk' \in K_o$; similarly one shows that $k^{-1} \in K_o$ and K_o is a closed subgroup of K.

Let $z \in Z$ and let $\{s_i\}$ be a net in T such that $z = \lim s_i z_o$ then clearly $zK_o \subset \psi^{-1}(\psi(z))$. If $k \in K$ and $\psi(zk) = \psi(z)$ then $\lim s_i \psi(z_o k) = \lim s_i \psi(z_o)$ i.e. $\psi(z_o k)$ and $\psi(z_o)$ are proximal points of X. But $\phi(\psi(z_o k) = \chi(z_o k) = \chi(z_o) = \phi(\psi(z_o))$ and ϕ is an almost periodic hence a distal extension. Thus $\psi(z_o k) = \psi(z_o)$ and $k \in K_o$, therefore $\psi^{-1}(\psi(z)) = zK_o$ and $(T,Z,K_o) \to (T,X)$ is a group extension.

Now suppose that $(T,X,L) \xrightarrow{\phi} (T,Y)$ is a group extension (and L acts freely). Let $\theta(k)$ be the unique element of L such that $\chi(z_o k) = x_o \ell$. Let $\lim t_i z_o = z_o k$, then

$$x_o = \psi(z_o) = \lim t_i \psi(z_o k^{-1}) = \lim t_i x \theta(k^{-1}) =$$

$$= \lim \psi(t_i z_o)\theta(k^{-1}) = x_o \theta(k)\theta(k^{-1}).$$

Thus $\theta(k)\theta(k^{-1}) = e$, and similarly for $k,k' \in K$ $\theta(k)\theta(k') = \theta(kk')$ and θ is a homomorphism. Clearly θ is onto L and $\ker \theta = K_o$. Finally if $z \in Z$ and $z = \lim s_i z_o$ then

$$\psi(zk) = \lim s_i \psi(z_o k) = \lim s_i \psi(z_o)\theta(k)$$

$$= \psi(z)\theta(k). //$$

Fix a minimal pointed flow (T,Y,y_o) and consider all pointed group extensions

$$\chi_\alpha : (T,Z_\alpha,K_\alpha,z_\alpha) \to (T,Y,y_o).$$

Let z_o be the element of $\prod_\alpha Z_\alpha$ such that $z_o(\alpha) = z_\alpha$ and let $Z =$

cls (Tz_o). Since our base points are always chosen so that $ux_o = x_o$ for a distinguished idempotent u in a fixed minimal ideal M of βT (Chapter I, section 4), it follows that $uz_o = z_o$ and hence (T,Z) is minimal (Proposition I.3.1.(2).). There is a natural homomorphism χ of Z onto Y defined by $\chi(z) = \chi_\alpha(z(\alpha))$. Let $K' = \prod_\alpha K_\alpha$ and let $K = \{k \in K' \mid z_o k \in Z\}$; it is easy to check that K is a closed subgroup of K and that

$$\chi\colon (T,Z,K,z_o) \to (T,Y,y_o)$$

is a <u>universal</u> <u>group</u> <u>extension</u> and hence also a <u>universal</u> <u>almost</u> <u>periodic</u> <u>extension</u> of (T,Y,y_o).

We note that since all our homomorphisms are pointed this universal extension is unique up to an isomorphism. Indeed if

$$\begin{array}{ccc} (T,Z,K,z_o) & \underset{\lambda'}{\overset{\lambda}{\rightleftarrows}} & (T,Z',K',z_o') \\ & {\scriptstyle\chi}\searrow \quad \swarrow{\scriptstyle\chi'} & \\ & (T,Y,y_o) & \end{array}$$

are two such flows then by the universality there exist maps λ and λ' as above. Now $\lambda'\circ\lambda$ is an endomorphism of Z onto itself such that $\lambda'\circ\lambda(z_o) = z_o$, hence $\lambda'\circ\lambda$ = identity and Z and Z' are isomorphic (over Y).

VIII.2. THE GENERALIZED (STRONG) BOHR COMPACTIFICATION

<u>2.1. PROPOSITION</u>: <u>A minimal flow (T,X) is a compactification flow for T iff X is a group extension of a proximal flow</u>.

<u>Proof</u>: Let (T,X) be a compactification flow (see the introduction to this chapter) and let K be the group of automorphisms of (T,X). Since X is regular $(T,X/K)$ is a proximal flow and (T,X,K) is a

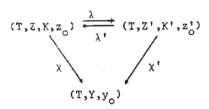

group extension of $(T,X/K)$.

Conversely if $(T,X,K) \xrightarrow{\chi} (T,Y)$ is a group extension and (T,Y) is proximal, then given $x,z \in X$ there exists a net $\{t_i\}$ in T such that $\lim t_i\chi(x) = \lim t_i\chi(z) = \chi(z)$.

We can assume that the limits, $z_1 = \lim t_i x$ and $z_2 = \lim t_i z$ exist and then there exist k_1 and k_2 in K such that $z_1 = zk_1$ and $z_2 = zk_2$. Now $\lim t_i(xk_1^{-1}k_2) = z_1k_1^{-1}k_2 = z_2$. Since also $\lim t_i z = z_2$, $xk_1^{-1}k_2$ is an automorphic image of x which is proximal to z, and X is regular. Moreover if ϕ is an automorphism of (T,X) and $x_o \in X$ then by I.3.3. x_o and $\phi(x_o)$ are distal, hence $\chi(x_o) = \chi(\phi(x_o))$ i.e., $\phi(x_o) = x_o k$ for some $k \in K$. By the minimality of (T,X), $\phi(x) = xk$ for every $x \in X$. So that K is the group of all automorphisms of (T,X).//

Motivated by this proposition we now say that a minimal flow (T,X) is a <u>strong compactification flow</u> if there exists a compact Hausdorff topological group K such that $(T,X,K) \to (T,X/K)$ is a group extension and such that $(T,X/K)$ is strongly proximal. We say that K is a <u>generalized strong compactification</u> of T.

Finally we have the following definitions. Let (T,π) $\{(T,\pi_S)\}$ be the universal minimal {strongly} proximal flow and let $(T,\pi^\#,\Sigma(T))$ $\{(T,\pi_S^\#,\Sigma_S(T))\}$ be the universal group extension of (T,π) $\{(T,\pi_S)\}$. Then $(T,\pi^\#)$ $\{(T,\pi_S^\#)\}$ is called the <u>universal</u> {<u>strong</u>} <u>compactification flow</u> of T and $\Sigma(T)$ $\{\Sigma_S(T)\}$ is the <u>generalized</u> {<u>strong</u>} <u>Bohr compactification</u> of T.

<u>2.2. EXAMPLE</u>: Let $\mathbb{G} = \mathbb{SL}(2,\mathbb{R})$, let $Y = \mathbb{P}^1$ be the projective line, the space of all lines through the origin in $\mathbb{R}^2$, and let X be the space of rays emanating from the origin which we will think of as the unit circle S^1. Using the linear action of $\mathbb{G}$ on $\mathbb{R}^2$ we obtain two minimal flows $(\mathbb{G},X)$ and $(\mathbb{G},Y)$. Letting $K = \{\sigma,\text{id.}\}$ where σ is the antipodal map and id is the identity map on X we

obtain the bitransformation group $(\mathbb{G},X,K)$ which is a group extension of the proximal flow $(\mathbb{G},Y)$. In other words, $\mathbb{Z}_2$ is a generalized compactification of $\mathbb{G}$.

Since X is a regular flow, it is isomorphic to a minimal ideal in its enveloping semigroup and we can easily identify the semigroup structure on X. Let $u = \lim_{b\to\infty}\begin{pmatrix}1 & b\\ 0 & 1\end{pmatrix}$ in the enveloping semigroup. Let $J = \{z \in S^1 \mid \operatorname{Im}(z) > 0\} \cup \{1\}$, and let $J' = S^1\setminus J$. Then it is easy to see that $u(J) = 1$ and $u(J') = -1$. For $z \in J$ and $g \in \mathbb{G}$ such that $g1 = z$, let $v_z = gu$ and note that this does not depend upon g. For $z \in J'$ let $v_z = \alpha v_{\alpha z}$ where $\alpha = \begin{pmatrix}-1 & 0\\ 0 & -1\end{pmatrix}$. Note that $v_1 = u$ and $v_{-1} = \alpha u$. It is easy to check that $\mathbb{G}u = \{v_z \mid z \in S^1\}$, $z \to v_z$ is a homeomorphism of S^1 into the enveloping semi-group of X, and $gv_z = v_{gz}$ for $g \in \mathbb{G}$. Thus $\mathbb{G}u$ is a minimal left ideal in this enveloping semigroup and $z \to v_z$ is a flow isomorphim. The semigroup structure on X is given by $zw = z$ if $w \in J$ and $zw = \sigma(z)$ if $w \in J'$. Thus J is the set of idempotents of X.

For any $z_0 = e^{i\theta}$ we could carry out this construction by replacing $\begin{pmatrix}1 & b\\ 0 & 1\end{pmatrix}$ by $\begin{pmatrix}\cos\theta & -\sin\theta\\ \sin\theta & \cos\theta\end{pmatrix}\begin{pmatrix}1 & b\\ 0 & 1\end{pmatrix}$. This gives a description of all minimal ideals. It is also easy to check that the map $p' \to p'u$ will not send idempotents to idempotents as p' ranges over some minimal ideal $\mathbb{G}u'$. Thus the natural flow isomorphism does not preserve the semigroup structure. (See the remark after proposition I.2.5).

VIII.3. THE GENERALIZED STRONG BOHR COMPACTIFICATION OF A CONNECTED SEMISIMPLE LIE GROUP WITH FINITE CENTER.

Let (T,Y) be a flow and let L be a compact topological group. A <u>cocycle</u> is a continuous map $\sigma\colon T\times Y \to L$ such that $\sigma(st,y) = \sigma(s,ty)\sigma(t,y)$ for $s,t \in T$ and $y \in Y$. This latter condition is equivalent to $(t,(y,\ell)) \to (ty,\sigma(t,y)\ell)$ is a flow on $Y\times L$.

3.1. LEMMA: Let $\chi: (T,X,L) \to (T,Y)$ be a group extension (with L acting freely) and let $\psi: Y \to X$ be a homeomorphism into X, such that $\chi \circ \psi$ is the identity on Y. Then X is homeomorphic to $Y \times L$ and there exists a cocycle $\sigma: T \times Y \to L$ such that (T,X) is isomorphic to the flow on $Y \times L$ induced by σ.

Proof: Define $\gamma: Y \times L \to X$ by $\gamma(y,\ell) = \psi(y)\ell$. Clearly a continuous map onto, it suffices to show it is one-to-one. Suppose $\psi(y)\ell = \psi(y')\ell'$. By applying χ we see that $y = y'$ and then $\ell = \ell'$ because L acts freely on X.

Define $(T, Y \times L)$ by $t(y,\ell) = \gamma^{-1}(t\gamma(y,\ell))$. For $y \in Y$ and $t \in T$ we have $\chi(t\psi(y)) = t\chi\psi(y) = ty$. Hence there exists a unique $\ell' \in L$ such that $t\psi(y) = \psi(ty)\ell'$. Denote $\ell' = \sigma(t,y)$; clearly σ is continuous and

$$t(y,\ell) = \gamma^{-1}(t\gamma(y,\ell)) = \gamma^{-1}(t(\psi(y)\ell)) =$$

$$= \gamma^{-1}((t\psi(y))\ell) = \gamma^{-1}(\psi(ty)\ell'\ell) = (ty,\ell'\ell)$$

$$= (ty, \sigma(t,y)\ell').$$

Since $(t,(y,\ell)) \to t(y,\ell)$ is a flow, σ is a cocycle.//

If $\chi: (T,X,L) \to (T,Y)$ is a group extension and a section $\psi: Y \to X$ as in lemma 2.1. exists we shal say that X is a cocycle extension of Y and denote $X \simeq Y \times_\sigma L$ where σ is the cocycle which corresponds to ψ.

Let $\mathbb{G}$ be a connected semisimple Lie group with finite center. We will describe its generalized strong Bohr compactification. Let $\mathbb{G} = \mathbb{K}\mathbb{A}\mathbb{N}$ be an Iwasawa decomposition for $\mathbb{G}$ (IV.1.5.). Let $\mathbb{S} = \mathbb{A}\mathbb{N}$ and let $\mathbb{M}$ be the centralizer of $\mathbb{A}$ in $\mathbb{K}$, then $\mathbb{P} = \mathbb{M}\mathbb{A}\mathbb{N}$ is the normalizer of $\mathbb{N}$ in $\mathbb{G}$ and also the normalizer of $\mathbb{S}$ in $\mathbb{G}$ (IV.3.1. and the following remark). Finally we recall that the homogeneous flow $(\mathbb{G}, \mathbb{G}/\mathbb{P})$ is isomorphic to the universal strongly proximal flow $(\mathbb{G},\pi_S(\mathbb{G}))$. (Theorem IV.3.2.)

3.2. LEMMA: *Let* $Y = \mathbb{G}/\mathbb{S}$, $y_o = \{\mathbb{S}\} \in Y$ *and let* $(\mathbb{G},X,L) \xrightarrow{\chi} (\mathbb{G},Y)$ *be a group extension. Then* X *is a cocyclic extension of* Y. *Moreover the cocycle* σ *can be chosen so that* $\sigma(k,y) = e$ *for every* $k \in \mathbb{K}$ *and* $y \in Y$, *and then* $\sigma(kg,y) = \sigma(g,y)$ *and* $\sigma(gk,y) = \sigma(g,ky)$ *for every* $k \in \mathbb{K}$, $g \in \mathbb{G}$ *and* $y \in Y$. *If* $s \in \mathbb{S}$ *and we put* $\lambda(s) = \sigma(s,y_o)$ *then* λ *is a homomorphism of* $\mathbb{S}$ *onto a dense subgroup of* L (*i.e.* $\mathbb{S} \xrightarrow{\lambda} L$ *is a compactification of* $\mathbb{S}$) *and* σ *is determined by* λ.

Proof: Consider the flow $(\mathbb{K},X)$ and note that $\mathbb{K}$ acts transitively and freely on Y. Pick $x_o \in \chi^{-1}(y_o)$ and define $\psi\colon Y \to X$ by $\psi(ky_o) = kx_o$. Lemma 2.1. applies and $(\mathbb{G},X,L)$ is isomorphic to $(\mathbb{G},Y \times_\sigma L,L)$ where $\sigma(g,y)$ is defined by $g\psi(y) = \psi(gy)\sigma(g,y)$. Since for $k \in \mathbb{K}$ and $y \in Y$ $k\psi(y) = \psi(ky)$, it follows that $\sigma(k,y) = e$. For $g \in \mathbb{G}$ $\sigma(kg,y) = \sigma(k,gy)\sigma(g,y) = \sigma(g,y)$ and $\sigma(gk,y) = \sigma(g,ky)\sigma(k,y) = \sigma(g,ky)$.

If $s_1,s_2 \in \mathbb{S}$ then

$$\lambda(s_1s_2) = \sigma(s_1s_2,y_o) = \sigma(s_1,s_2y_o)\sigma(s_2,y_o) =$$

$$= \sigma(s_1,y_o)\sigma(s_2,y_o) = \lambda(s_1)\lambda(s_2).$$

Thus λ is a homomorphism of $\mathbb{S}$ into L. Moreover since $(\mathbb{G},Y \times_\sigma L)$ is minimal and for $g = ks$

$$g(y_o,e) = (gy_o,\sigma(g,y_o)) = (gy_o,\sigma(ks,y_o)) =$$

$$= (gy_o,\sigma(s,y_o)) = (gy_o,\lambda(s)),$$

it follows that $\lambda(\mathbb{S})$ is dense in L.

Finally if $g = ks$ and $y = k'y_o$ then for some $k_1 \in \mathbb{K}$ and

$s_1 \in \mathbb{S}$, $sk' = k_1 s_1$ and

$$\sigma(g,y) = \sigma(ks,k'y_o) = \sigma(sk',y_o) = \sigma(k_1 s_1,y_o) = \lambda(s_1)$$

and σ is determined by λ.//

3.3. LEMMA: Let $\lambda\colon \mathbb{S} \to L$ be a compactification of $\mathbb{S}$, then $\mathbb{N} = [\mathbb{S},\mathbb{S}] \subset \ker \lambda$. Define $\eta\colon \mathbb{G} \to \mathbb{S}$ by $\eta(g) = s$ if $g = ks$ and define σ by $\sigma(g,y) = \sigma(g,ky_o) = \lambda(\eta(gk))$, then σ is a cocycle and the flow $(\mathbb{G},Y \times_\sigma L)$ is minimal. Thus there is one to one correspondence between group extensions of $(\mathbb{G},Y)$ and compactifications of $\mathbb{A}$.

Proof: Let $\lambda\colon \mathbb{S} \to L$ be a compactification of $\mathbb{S}$ then L is a compact solvable group and there are sufficiently many compact Lie group factors of L. Each of them is compact and solvable hence abelian. Thus L itself is abelian and $\mathbb{N} = [\mathbb{S},\mathbb{S}] \subset \ker \lambda$.

Let $g_1,g_2 \in \mathbb{G}$ and $y = ky_o \in Y$ then there exist k_i and s_i in $\mathbb{K}$ and $\mathbb{S}$ respectively $(i = 1,2)$ such that $g_2 k = k_2 s_2$ and $g_1 k_2 = k_1 s_1$. Then

$$\sigma(g_1 g_2,y) = \lambda(\eta(g_1 g_2 k)) = \lambda(\eta(k_1 s_1 s_2)) = \lambda(s_1)\lambda(s_2)$$

and

$$\sigma(g_1,g_2 y)\sigma(g_2,y) = \sigma(g_1,k_2 s_2 y_o)\sigma(g_2,y) =$$

$$= \sigma(g_1,k_2 y_o)\sigma(g_2,y) = \lambda(\eta(g_1 k_2))\lambda(\eta(g_2 k)) =$$

$$= \lambda(s_1)\lambda(s_2).$$

So that σ is a cocycle. Finally because $\lambda(\mathbb{S})$ is dense in L and $g(y_o,e) = (ky_o,\lambda(s))$ where $g = ks$, it follows that $(\mathbb{G},Y \times_\sigma L)$ is minimal (See Lemma II.1.2.).//

3.4. LEMMA: Let $\mathbb{M}$ act on Y by $(g\mathbf{S},m) \to g\mathbf{S}m = gm\mathbf{S}$, then $(\mathbb{G},Y,\mathbb{M})$ is a group extension of $(\mathbb{G},\pi_S(\mathbb{G}))$. Moreover if $Y \underset{\sigma}{\times} L$ is any group extension of Y (see lemma 3.2.) then the action of $\mathbb{M}$ on $Y \underset{\sigma}{\times} L$ defined by $(y,\ell)m = (ym,\ell)$ commutes with the action of $\mathbb{G}$ on $Y \underset{\sigma}{\times} L$. Thus $(\mathbb{G},Y \underset{\sigma}{\times} L, L\times\mathbb{M}) \to (\mathbb{G},\pi_S(\mathbb{G}))$ is a group extension.

Proof: Clearly $(\mathbb{G},Y,\mathbb{M}) \to (\mathbb{G},\pi_S(\mathbb{G}))$ is a group extension. Now for $(y,\ell) \in Y \underset{\sigma}{\times} L$ and $g \in \mathbb{G}$ and $m \in \mathbb{M}$ we have $y = ky_o$ for some $k \in \mathbb{K}$ and

$$g[(y,\ell)m] = g(ym,\ell) = (gym,\sigma(g,ym)\ell)$$

$$= (gym,\sigma(g,kmy_o)\ell) = (gym,\lambda(\eta(gkm))\ell)$$

and

$$[g(y,\ell)]m = (gym,\sigma(g,y)\ell) = (gym,\sigma(g,ky_o)\ell)$$

$$= (gym,\lambda(\eta(gk))\ell).$$

Thus to show that $g[(y,\ell)m] = [g(y,\ell)]m$ it suffices to show that $\lambda(\eta(gkm)) = \lambda(\eta(gk))$. Let $gk = k_1an$ where $k_1 \in \mathbb{K}$, $a \in \mathbb{A}$ and $n \in \mathbb{N}$. Then $gkm = k_1anm = k_1amn' = k_1man'$ where $n' \in \mathbb{N}$, so that $\eta(gkm) = an'$. Now $gk = k_1an$ and $\eta(gk) = an$; since $\mathbb{N} \subset \ker\lambda$ $\lambda(\eta(gk)) = \lambda(\eta(gkm)) = \lambda(a)$.//

3.5. LEMMA: Let $\lambda_o\colon \mathbb{A} \to \Gamma(\mathbb{A})$ be the Bohr compactification of the abelian group $\mathbb{A}$. Let σ_o be the corresponding cocycle and let $(\mathbb{G},Y^{\#},L)$ be the universal group extension of $(\mathbb{G},Y)$. Then L is isomorphic to $\Gamma(\mathbb{A})$ and $Y^{\#} \cong Y \underset{\sigma}{\times} \Gamma(\mathbb{A})$.

Proof: By lemmas 3.2 and 3.3 $Y^{\#} \cong Y \underset{\sigma}{\times} L$ where σ corresponds to a compactification $\mathbb{A} \xrightarrow{\lambda} L$. Let $\theta\colon \mathbb{A} \to K$ be a compactification of $\mathbb{A}$ and let ρ be the corresponding cocycle then by the universality of

$Y^{\#}$ there exists a homomorphism

$$Y \underset{\sigma}{\times} L \xrightarrow{\alpha} Y \underset{\rho}{\times} K$$

such that $\alpha(y_o,e) = (y_o,e)$. (This is so because every group extension of Y, say $X \xrightarrow{\phi} Y$ is a distal extension and hence $ux_o = x_o$ implies $ux = x$ for every $x \in \phi^{-1}(y_o)$ (proposition I 4.1.(3).) so that we can choose (y_o,e) to be the base point).

It follows by lemma 1.1. that K is a homomorphic image of L, say $L \xrightarrow{\beta} K$ and the diagram

$$\begin{array}{ccc} \mathbb{A} & \xrightarrow{\lambda} & L \\ & \searrow^{\theta} & \downarrow \beta \\ & & K \end{array}$$

is commutative.

(If $a \in \mathbb{A}$ then $\alpha(a(y_o,e)) = \alpha(y_o,\lambda(a)) = \alpha((y_o,e)\lambda(a)) = (y_o,e)\beta(\lambda(a))$, and on the other hand $a(\alpha(y_o,e)) = a(y_o,e) = (y_o,\theta(a)) = (y_o,e)\theta(a)$. So that $\theta(a) = \beta(\lambda(a))$).

Thus $L \cong \Gamma(A)$ is the Bohr compactification of $\mathbb{A}$. //

Now let $(\mathbb{G},\pi_S^{\#},\Sigma_S(\mathbb{G}))$ be the universal compactification flow of $\mathbb{G}$ (i.e., the universal group extension of $\pi_S(\mathbb{G})$). Then since by lemma 3.4. $(\mathbb{G},Y,\mathbb{M})$ is a group extension of $(\mathbb{G},\pi_S)$ it follows that Y is a factor of $\pi_S^{\#}$; hence by lemma 1.1. $\pi_S^{\#}$ is a group extension of Y. On the other hand by lemma 3.4. every group extension of Y is also a group extension of π_S and hence it is a factor of $\pi_S^{\#}$. Thus $\pi_S^{\#}$ is the universal group extension of Y; i.e., by lemma 3.5.

$$(\mathbb{G},\pi_S^{\#},\Sigma(\mathbb{G})) \cong (\mathbb{G},Y \underset{\sigma_o}{\times} \Gamma(\mathbb{A}),\Gamma(\mathbb{A})\times\mathbb{M})$$

Thus we have proved:

<u>3.6. THEOREM</u>: <u>The generalized strong Bohr compactification</u> $\Sigma(\mathbb{G})$ <u>of</u>

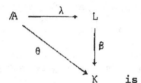

a connected semisimple Lie group $\mathbb{G}$ with finite center, is isomorphic to $\Gamma(\mathbb{A})\times\mathbb{M}$ where $\mathbb{G} = \mathbb{K}\mathbb{A}\mathbb{N}$ is an Iwasawa decomposition of $\mathbb{G}$, $\Gamma(\mathbb{A})$ is the Bohr compactification of $\mathbb{A}$ and $\mathbb{M}$ is the centralizer of $\mathbb{A}$ in $\mathbb{K}$.

3.7. EXAMPLE: Let $\mathbb{G} = \mathbb{SL}(2,\mathbb{R})$ then $\mathbb{K} = \left\{\begin{pmatrix} \cos\theta & -\sin\theta \\ \sin\theta & \cos\theta\end{pmatrix} \middle| 0 \le \theta \le 2\pi\right\}$, $\mathbb{A} = \left\{\begin{pmatrix} a & 0 \\ 0 & a^{-1}\end{pmatrix} \middle| a > 0\right\}$, $\mathbb{N} = \left\{\begin{pmatrix} 1 & b \\ 0 & 1\end{pmatrix} \middle| b \in \mathbb{R}\right\}$, $\mathbb{M} = \left\{\begin{pmatrix} 1 & 0 \\ 0 & 1\end{pmatrix}, \begin{pmatrix} -1 & 0 \\ 0 & -1\end{pmatrix}\right\}$ and for $g = \begin{pmatrix} x & y \\ z & w\end{pmatrix} \in \mathbb{G}$, the Iwasawa decomposition is given by

$$\begin{pmatrix} x & y \\ z & w\end{pmatrix} = \begin{pmatrix} \cos\theta & -\sin\theta \\ \sin\theta & \cos\theta\end{pmatrix}\begin{pmatrix} a & 0 \\ 0 & a^{-1}\end{pmatrix}\begin{pmatrix} 1 & b \\ 0 & 1\end{pmatrix}$$

where $a = \sqrt{x^2+y^2}$, $x = a\cos\theta$, $z = a\sin\theta$ and $b = (y+a^{-1}\sin\theta)/x$.

The universal strongly proximal flow $\pi_S = \mathbb{G}/\mathbb{M}\mathbb{A}\mathbb{N}$ is the projective line $\mathbb{P}'$ (IV.4.1.) and $(\mathbb{G},Y,\mathbb{M}) \to (\mathbb{G},\pi_S)$ where $Y = \mathbb{G}/\mathbb{S} = \left\{\begin{pmatrix}\cos\theta \\ \sin\theta\end{pmatrix} \middle| 0 \le \theta \le 2\pi\right\}$ is the group extension of example 2.2.

For $r \in \mathbb{R}$, $r \ne 0$ we take $\lambda_r\colon \mathbb{A} \to S' = \{z \in \mathbb{C} \mid |z| = 1\}$ to be the homomorphism

$$\lambda_r\begin{pmatrix} a & 0 \\ 0 & a^{-1}\end{pmatrix} = e^{2\pi i r \log a}.$$

If we let $\begin{pmatrix}1\\0\end{pmatrix} = v_0 \in Y$, $\begin{pmatrix}\cos\theta\\ \sin\theta\end{pmatrix} = v \in Y$, $\begin{pmatrix}\cos\theta & -\sin\theta \\ \sin\theta & \cos\theta\end{pmatrix} = k \in \mathbb{K}$ and $g \in \mathbb{G}$ then the cocycle σ_r which corresponds to λ_r satisfies

$$\sigma_r(g,v) = \sigma_r(g,kv_0) = \lambda_r(\eta(gk)) = \lambda_r\begin{pmatrix} \|gv\| & 0 \\ 0 & \|gv\|^{-1}\end{pmatrix} =$$

$$= e^{2\pi i r \log \|gv\|},$$

where $\left\|\begin{pmatrix}x\\y\end{pmatrix}\right\| = \sqrt{x^2+y^2}$. This gives a minimal flow on the torus $Y \times S'$

$$g(v,z) = \left(\frac{gv}{\|gv\|}, e^{2\pi i r \log \|gv\|} z \right).$$

Since the Bohr compactification of $\mathbb{A}$ is the inverse limit of all the homomorphisms λ_r $(r \in \mathbb{R})$, $\pi_S^{\#}$ is the inverse limit of the corresponding flows. The generalized strong Bohr compactification of $\mathbb{SL}(2,\mathbb{R})$ is $\Gamma(\mathbb{R}) \times \mathbb{Z}_2$ where $\Gamma(\mathbb{R})$ is the Bohr compactification of the real line and $\mathbb{Z}_2$ is the cyclic group of two elements.

CHAPTER IX

THE GENERALIZED BOHR COMPACTIFICATION, II

For convenience we assume from now through the rest of these notes that T is a discrete group. What follows is the work of many people but the main ideas and constructions are due to R. Ellis. He introduced the τ-topologies and discovered the relation between these topologies and the notion of an almost periodic extension. Among other things he utilized this relations to give an elegant "algebraic" proof of the Furstenberg structure theorem for distal flows [13] and then he was able to complete the proof of Veech's theorem about the structure of point distal flows [41, 10].

In [12], generalizing a construction which was introduced in [21], the circle operation is defined and it was shown there that this operation can yield a convenient and elegant way of presenting the τ-topologies. Via this presentation new results about the τ-topologies were proved and the proofs of many old results were simplified.

It also turned out that the "circle operation" is very useful in the construction certain extensions, RIC-extension, which is a generalization of the notion of a distal extension.

All these tools being put together enabled the authors of [12] to present a structure theory for PI-flows; i.e. flows which can be constructed by iterating proximal and almost periodic extensions. This theory includes among other things the Furstenberg structure theorem for distal flows [13], structure theorems for flows possesing a point with a countable proximal cell and for flows in which the proximal relation is an equivalence relation [12], and a structure theorem for flows in whose enveloping semigroup there are only finitely many minimal ideals (recent work of the author).

In the next two chapters we describe most of the work which was done in [12], which in turn is based mostly on works of Ellis (in particular [10]). We used this opportunity to represent the τ-topology in a way which is based entirely on the circle operation. This is one of the reasons why we give here proofs for the central theorem 4.2 and its relative analogue theorem X.2.1., which are different from the ones which are given in [12]. As a matter of fact the proofs which are given here to these theorems where the original proofs discovered by the authors of [12]. Another novelty of our representation is that we dispense all together with $\tau(a)$-topologies (see [9]) and work only with the τ-topology on G.

In this chapter we describe the first step in the construction of the universal PI-tower, namely the maximal almost periodic extension of the universal minimal proximal flow or in other words the universal compactification flow of T.

IX.1. THE τ-TOPOLOGY ON G.

Let (T,X) be a flow and let 2^X be the set of all closed non-empty subsets of X. For a finite collection $\{U_1,\cdots,U_n\}$ of open subsets of X set

$$\langle U_1,\cdots,U_n\rangle = \{A \in 2^X \mid A \subset \bigcup_{i=1}^{n} U_i \text{ and } A \cap U_i \neq \emptyset \quad \forall i\}.$$

Then the collection of all $\langle U_1,\cdots,U_n\rangle$ is a basis for a compact Hausdorff topology on 2^X. In this topology $\lim_{i\in\Omega} A_i = A$ where $\{A_i\}_{i\in\Omega}$ is a net in 2^X and $A \in 2^X$ implies

$$A = \{\lim_{j\in\Omega'} x_j \mid x_j \in A_j \quad \forall j \in \Omega' \text{ and } \{A_j\}_{j\in\Omega'} \text{ is a subnet of } \{A_i\}_{i\in\Omega}\}.$$

With this topology 2^X is metric iff X is metric. There is a natural flow on 2^X induced by (T,X), namely $(t,A) \to tA = \{tx \mid x \in A\}$.

We say that a minimal set of the flow $(T,2^X)$ is a quasi-factor of (T,X). (See [21] for more details)

As was explained in chapter I section 2. there is an action of βT, the Stone Chêch compactification of the (discrete) group T on 2^X. We recall that this action is the extension of the map $\psi(t) = tA$ from T to 2^X to a map $\tilde{\psi}$ of βT into 2^X. If $p \in \beta T$ and $p = \lim t_i$ for some net $\{t_i\}$ in T then $\tilde{\psi}(p) = \lim t_i A$.

Now (T,X) itself is a flow and there is an action of βT on X as well. We denote this action as usual by $(p,x) \to px$. This makes the notation pA ambiguous; we can think of pA as the, not necessarily closed, subset $\{px \mid x \in A\}$ of X or as the closed subset $\tilde{\psi}(p) = \lim t_i A$. We shall adopt the following notation $pA = \{px \mid x \in A\}$ and $p \circ A = \tilde{\psi}(p)$. We have

1.1. LEMMA: *For* $A \in 2^X$ *and* $p,q \in \beta T$

(1) $p \circ A$ *is the set of all points* $x \in X$ *such that there exist nets* $\{x_i\}$ *in* A *and* $\{t_i\}$ *in* T *for which* $\lim t_i = p$ *and* $\lim t_i x_i = x$.

(2) $pA \subset p \circ A$.

(3) $p \circ (q \circ A) = (pq) \circ A$.

Proof: (1) follows from the definition of $p \circ A$ as $\lim t_i A$ where $\{t_i\}$ is any net in T which converges to p. Now (2) follows by noting that for a fixed $x \in A$, $\lim t_i x = px$ whenever $\lim t_i = p$.

Statement (3) is just the general fact that βT acts on every flow as a semigroup. //

If $A \subset X$ is, not necessarily closed, subset of X we define $p \circ A = p \circ \bar{A}$. Clearly the characterization of $p \circ A$ given in lemma 1.1. (1), is still valid.

We recall that G is the subgroup uM of M (See chapter I).

1.2. PROPOSITION: *The operation* $A \to (u \circ A) \cap G$, *for subsets of* G, *defines a closure operator on* G.

Proof: First $A = uA = uA \cap G = (u \circ A) \cap G$. Clearly $A \subset B$ implies $(u \circ A) \cap G \subset (u \circ B) \cap G$. Next $u \circ [(u \circ A) \cap G] \cap G \subset (u \circ u \circ A) \cap G = (u^2 \circ A) \cap G = (u \circ A) \cap G$. Finally $[u \circ (A \cup B)] \cap G = [(u \circ A) \cup (u \circ B)] \cap G = [(u \circ A) \cap G] \cup [(u \circ B) \cap G]$.
//

We denote this operation by $\bar{A}^{\tau}$ or $cls_{\tau}(A)$ and call the topology induced on G by it the τ-*topology* (This coincide with R. Ellis' τ-topology [9].)

1.3. LEMMA: $cls_{\tau} A = u(u \circ A)$.

Proof: $(u \circ A) \cap G \subset u(u \circ A)$ which is contained in both G and $u \circ (u \circ A) = u \circ A$.
//

If $v \in J$ the operator $A \to (v \circ A) \cap vG$ defines a closure operator on the group $vG = vM$ and we have the following:

1.4. LEMMA: *The map* $\gamma \to v\gamma$ *of* $G = uM$ *onto* $vG = vM$ *is a* τ-*isomorphism*.

Proof: We already know that this map is an isomorphism of groups (proposition I.2.3.(3).). Let $A \subset G$ be τ-closed then:

$$A = uvA = uvvA \subset u(v \circ vA) \subset u(v \circ v \circ A) = u(v \circ A) = uu(v \circ A)$$
$$\subset u(u \circ v \circ A) = u(u \circ A) = A.$$

Thus $u(v \circ vA) = A$ and hence also $v(v \circ vA) = vA$; i.e. vA is τ-closed in vG.
//

We have the following relation between the original topology on M and the τ-topology on G.

1.5. LEMMA: *Let* $\{\alpha_i\}$ *be a net in* G. *If in* M $\lim \alpha_i = p$ *then*

τ-lim α_i = up. <u>Thus the τ-topology on G is weaker than the topology induced on it from</u> M.

<u>Proof</u>: Let $A_j = \{\alpha_i \mid i \geq j\}$ then $uA_j = A_j$ and hence $p \in \bar{A}_j \subset u \circ A_j$. It follows that $up \in u(u \circ A_j) = cls_\tau(A_j)$. Thus up is a τ-limit of a subnet of $\{\alpha_i\}$. Since this is true also for every subnet of $\{\alpha_i\}$ we have τ-lim α_i = up. //

<u>1.6. PROPOSITION</u>: G <u>with the τ-topology is a compact T_1 space</u>.

<u>Proof</u>: Clearly $u \circ \{\alpha\} = \{\alpha\} = cls_\tau(\{\alpha\})$ for every $\alpha \in G$, hence G is T_1. Let $\{\alpha_i\}$ be a net in G then by the compactness of M there exists a subnet $\{\alpha_{i_j}\}$ and an element $p \in M$ such that $p = \lim \alpha_{i_j}$. By lemma 1.5. τ-lim $\alpha_{i_j} = up \in G$ and hence G is compact. //

<u>1.7. PROPOSITION</u>: <u>Both left and right multiplication by a fixed element of</u> G <u>are homeomorphisms with respect to the τ-topology</u>.

<u>Proof</u>: Let $\alpha \in G$ and let A be a τ-closed subset of G. It suffices to show that $cls_\tau(A\alpha) = A\alpha$ and $cls_\tau(\alpha A) = \alpha A$. Now

$$cls_\tau(A\alpha) = u(u \circ A\alpha) = u[(u \circ A)\alpha] =$$

$$= [u(u \circ A)]\alpha = A\alpha.$$

For the other equality we first observe that $(\alpha \circ A) \cap G = \alpha A$ because $\beta \in (\alpha \circ A) \cap G$ implies that $\alpha^{-1}\beta \in (u \circ A) \cap G = A$; i.e., $\beta \in \alpha A$. Also note that (as in lemma 1.3.) $u(\alpha \circ A) = (\alpha \circ A) \cap G$. The following chain of inclusions completes the proof.

$$\alpha A \subset cls_\tau(\alpha A) = u(u \circ \alpha A) \subset u(u \circ \alpha \circ A) =$$

$$= u(\alpha \circ A) = (\alpha \circ A) \cap G = \alpha A.$$

//

Let F be a τ-closed subgroup of G. Then with the induced τ-topology, F is a compact T_1 group with continuous right and left multiplication. Following Ellis we proceed to define the closed subgroup $H(F)$ of F and to show that $H(F)$ is the minimal closed subgroup with the property that $F/H(F)$ with the quotient topology is a compact Hausdorff topological group. [9].

Let N be a neighbourhood filter for the τ-topology on G at u and let N_F be the neigbourhood filter for the relative τ-topology on F at u; i.e.,

$$N_F = \{V \cap F \mid V \in N\}.$$

1.8. LEMMA: *Let* $D \subset F$ *then*

$$\mathrm{cls}_\tau D = \bigcap \{V^{-1}D \mid V \in N_F\}.$$

Proof: Let $x \in \mathrm{cls}_\tau D$ and let $V \in N_F$. Then Vx is a neigbourhood of x and $Vx \cap D \neq \emptyset$ or $x \in V^{-1}D$. Now suppose $x \in V^{-1}D$ for all $V \in N_F$ and let W be a neigbourhood of x in F. There exists $V \in N_F$ such that $Vx \subset W$. Since $x \in V^{-1}D$, $Vx \cap D \neq \emptyset$ and $W \cap D \neq \emptyset$. //

1.9. THEOREM: *Let*

$$H(F) = \bigcap \{\mathrm{cls}_\tau V \mid V \in N_F\}$$

then:

(1) $H(F)$ *is a* τ-*closed normal subgroup of* F. *Moreover* $H(F)$ *is invariant under all topological automorphisms of* F.

(2) $F/H(F)$ *with the quotient topology is a compact Hausdorff topological group*.

(3) *Let* K *be a* τ-*closed subgroup of* F *then* F/K *is a Hausdorff space iff* $K \supset H(F)$.

Proof: (1) Let V be an open element of N_F. We will first show that $\overline{V}^\tau H(F) \subset \overline{V}^\tau$. Let $x \in H(F)$ and $y \in \overline{V}^\tau$. For all $W \in N_F$,

$Wy \cap V \neq \emptyset$ and $wy \in V$ for some $w \in W$. Since V is open there exists $U \in N_F$ such that $wyU \subset V$. Now $wy\overline{U}^\tau \subset \overline{wyU}^\tau \subset \overline{V}^\tau$ and since $x \in \overline{U}^\tau$, $yx \in W^{-1}\overline{V}^\tau$. By lemma 1.8. $yx \in \overline{V}^\tau$ and $\overline{V}^\tau H(F) \subset \overline{V}^\tau$.

Clearly $\overline{V}^\tau H(F) \subset \overline{V}^\tau$ for every open V in N_F implies $H(F)^2 \subset H(F)$ and $H(F)$ is a semigroup. If we can conclude that for every $x \in H(F)$ the τ-closed subsemigroup $xH(F)$ contains an idempotent, then $H(F)$ is a subgroup of F because the identity is the only idempotent in the group F. Lemma I.2.2. does not quite apply because with the τ-topology F is not Hausdorff. However, in the proof of I.2.2. we only used the Hausdorff condition to show that Sx was closed when S was closed which holds in $H(F)$ because $y \to yx$ us a τ-homeomorphism of F onto F and $H(F)$ is a τ-closed subset. Thus $H(F)$ is a subgroup. Finally if η is a topological automorphism of F and $V \in N_F$ then there exists $U \in N_F$ such that $\eta(U) \subset V$. Then $\eta(\overline{U}^\tau) \subset \overline{V}^\tau$ and $\eta(H(F)) \subset \overline{V}^\tau$ for all $V \in N_F$; i.e., $\eta(H(F)) \subset H(F)$. In particular, $H(F)$ is normal in F.

(2) Suppose $xH(F) \neq yH(F)$ then $yx^{-1} \notin H(F)$ and there exists $V \in N_F$ such that $yx^{-1} \notin \overline{V}^\tau$. There exists $W \in N_F$ such that $yx^{-1} \notin W^{-1}\overline{V}^\tau$ (lemma 1.8.) or $Wyx^{-1} \cap \overline{V}^\tau = \emptyset$. Since $\overline{V}^\tau H(F) \subset \overline{V}^\tau$, we have $Wyx^{-1}H(F) \cap \overline{V}^\tau H(F) = \emptyset$ or $WyH(F) \cap VxH(F) = \emptyset$. Hence $F/H(F)$ is Hausdorff. Since the multiplication in $F/H(F)$ is continuous in each variable separately, Ellis' joint continuity theorem [8] implies that $F/H(F)$ is a topological group.

(3) Let $x \in H(F)$, for every $V \in N_F$, $x \in \overline{V}^\tau$. Hence by lemma 1.8. $x \in V^{-1}V$ or $Vx \cap V \neq \emptyset$. Let $x_V \in Vx \cap V$, then the net $\{x_V\}_{V \in N_F}$ converges both to x and to u. Hence in the space F/K the net $\{x_V K\}$ converges to both K and xK. If G/K is Hausdorff $K = xK$ and $x \in K$ i.e. $H \subset K$. The converse follows from (2). //

We next prove thru technical lemmas about the τ-topologies which we shall need later on.

1.10. LEMMA: *Let* K *and* L *be two* τ-*closed non-empty subsets of* G. *Then* KL *is* τ-*closed*.

Proof: Let $\alpha \in (u \circ KL) \cap G$ then $\alpha = \lim t_i \kappa_i \lambda_i$ where $\kappa_i \in K$ and $\lambda_i \in L$ and $\lim t_i = u$. We can assume that $\lim t_i \kappa_i = p \in M$ exists and then $\alpha \in p \circ L$. Applying up^{-1} we have $up^{-1}\alpha \in (u \circ L) \cap G = L$ and thus $\alpha \in upL$. Since $up \in u(u \circ K) = K$, $\alpha \in KL$. //

1.11. LEMMA: *Let* F *be a filter base of* τ-*closed subsets of* G. *Let* K *be a non-empty* τ-*closed subset of* G. *Then*

$$(\cap \{L \mid L \in F\})K = \cap \{LK \mid L \in F\}.$$

Proof: Since G is τ-compact $\cap \{L \mid L \in F\}$ is not empty, and it is clear that the left side is contained in the right. Let $\beta \in \cap \{LK \mid L \in F\}$. For each $L \in F$ let λ_L and κ_L be elements of L and K such that $\beta = \lambda_L \kappa_L$. Because F is a filter base $\{\lambda_L\}$ and $\{\kappa_L\}$ are nets. Let $\{\lambda_i\}$ be a subnet $\{\lambda_L\}$ converging to $p \in M$. Put $\alpha = up$, then by lemma 1.5. $\alpha = \tau\text{-}\lim \lambda_i$ and $\alpha \in L$ for each L. Now

$$\beta = \lim \lambda_i \kappa_i \in p \circ K$$

and thus $\alpha^{-1}\beta = up^{-1}\beta \in (u \circ K) \cap G = K$ i.e. $\beta \in \alpha K \subset (\cap \{L \mid L \in F\})K$. //

1.12. LEMMA: *Let* F *be a* τ-*closed subgroup of* T.

(1) *If* $\alpha \in F$, *then* $\alpha \circ F = u \circ F$.

(2) *Let* $p, q \in M$. *Then* $p \circ F = q \circ F$ *iff* $q \in p \circ F$.

Proof: (1) Since $F = \alpha F \subset \alpha \circ F$, $u \circ F \subset \alpha \circ F$. Likewise $u \circ F \subset \alpha^{-1} \circ F$ which implies that $\alpha \circ F \subset u \circ F$.

(2) Clearly $p \circ F = q \circ F$ implies that $q \in p \circ F$. Suppose $q \in p \circ F$ and as usual with $q = \lim t_i \alpha_i$ where $p = \lim t_i$ and $\alpha_i \in F$. Thus $q \circ F = \lim t_i \alpha_i \circ F$ and $p \circ F = p \circ u \circ F = \lim t_i(u \circ F)$. Since $t_i \alpha_i \circ F = t_i(\alpha_i \circ F) = t_i(u \circ F)$, $p \circ F = q \circ F$. //

IX.2. MORE ABOUT GROUP EXTENSIONS AND ALMOST PERIODIC EXTENSIONS.

We prove in this section a very useful characterization of almost periodic extensions based on the Ellis groups and H(F). The Ellis groups are all τ-closed. To see this let $p \in u \circ \mathcal{G}(X,x_0) \cap G$. Then $p = \lim t_i \alpha_i$ when $t_i \to u$ and $\alpha_i \in \mathcal{G}(X,x_0)$ from which it follows that $px_0 = \lim t_i \alpha_i x_o = ux_o = x_o$ and $p \in \mathcal{G}(X,x_0)$. Thus $H(\mathcal{G}(X,x_0))$ and Theorem 1.9. are at our disposal.

<u>2.1. THEOREM</u>: <u>Let</u> $(X,x_0) \xrightarrow{\phi} (Y,y_0)$ <u>be a distal homomorphism of minimal flows. Let</u> $\mathcal{G}(Y,y_0) = F$ <u>and</u> $\mathcal{G}(X,x_0) = A$.

(1) <u>If</u> A <u>is normal in</u> F <u>then every element</u> $\beta \in F$ <u>defines an automorphism</u> $\kappa(\beta)$ <u>of</u> X <u>such that</u> $\phi \circ \kappa(\beta) = \phi$. <u>The map</u> κ <u>from</u> F <u>into the group of automorphisms of</u> (T,X) <u>is a group homomorphism with kernel</u> A.

(2) ϕ <u>is a group extension</u>

$$(T,X,x_0,K) \xrightarrow{\phi} (T,Y,y_0)$$

<u>iff</u> A <u>is normal in</u> F <u>and</u> $A \supset H(F)$. <u>In that case</u> $\kappa(\beta) \in K$ <u>for every</u> $\beta \in F$ <u>and the map</u> $\kappa: F \to K$ <u>is a</u> τ<u>-continuous homomorphisms onto with kernel</u> A. <u>Moreover the natural map</u> $\tilde{\kappa}$: $F/A \to K$ <u>is a topological isomorphism</u>.

(3) <u>If</u> $A = H(F)$ <u>then the group extension</u> $(T,X,x_0,F/A) \to (T,Y,y_0)$ <u>is the universal group extension of</u> Y.

(4) ϕ <u>is an almost periodic extension iff</u> $A \supset H(F)$.

Proof: (1) For $p \in M$ and $\beta \in F$ define $(px_0)\kappa(\beta) = p\beta x_0$. If $px_0 = qx_0$ then $up^{-1}q \in A$ and $u(p\beta)^{-1}q\beta = \beta^{-1}(up^{-1}q)\beta \in \beta^{-1}A\beta = A$. Thus $up\beta x_0 = up\beta(u(p\beta)^{-1}q\beta)x_0 = uq\beta x_0$ and $p\beta x_0$ and $q\beta x_0$ are proximal. Since $\phi(q\beta x_0) = q\phi(\beta x_0) = qy_0 = py_0 = p\phi(\beta x_0) = \phi(p\beta x_0)$ these points are also distal and hence $p\beta x_0 = q\beta x_0$. Thus $\kappa(\beta)$ is well defined it is clearly a continuous automorphism of (T,X) with $\phi\circ\kappa(\beta) = \phi$. It is also clear that κ is a homomorphism with kernel A.

(2) Suppose first that A is normal in F and that $A \supset H(F)$. By theorem 1.9. F/A with the quotient τ-topology is a compact Hausdorff topological group. By (1) we have an action of F/A on X defined by $x(\beta A) = x\kappa(\beta)$. To show that this action is jointly continuous it suffices, by Ellis' joint continuity theorem to show that if $\{\beta_i\}$ is a net in F which is τ-convergent to β in F then $\lim x\kappa(\beta_i) = x\kappa(\beta)$ for every $x \in X$. By the definition of τ it suffices to show that for every $q \in M$ $\lim q\beta_i x_0 = q\beta x_0$. Let $r = \lim q\beta_i$ in M and let $v \in J$ such that $vr = r$. Since $rx_0 \in \phi^{-1}(qy_0)$ and since ϕ is a distal extension it follows that $vq\beta x_0 = q\beta x_0$. By lemmas 1.4. and 1.5. τ-lim $uq\beta_i = ur$ and τ-lim $ur^{-1}q\beta_i = u$. On the other hand τ-lim $ur^{-1}q\beta_i = ur^{-1}q\beta$ and since F/A is Hausdorff $uA = ur^{-1}q\beta A$ or $urx_0 = uq\beta x_0$. Finally multiplying by v we have $rx_0 = q\beta x_0$. Thus the action of F/A on X is jointly continuous and by I.4.1.(3) X/(F/A) is isomorphic to Y. Therefore $(T,X,x_0,F/A) \xrightarrow{\phi} (T,Y,y_0)$ is a group extension.

Conversely suppose that $(T,X,x_0,\kappa) \xrightarrow{\phi} (T,Y,y_0)$ is a group extension. Let $\alpha \in A$ and $\beta \in F$, then $\phi(\beta x_0) = \beta\phi(x_0) = \beta y_0 = y_0$. Therefore there exists $k \in K$ for which $\beta x_0 = x_0 k$. Now $\beta^{-1}\alpha\beta x_0 = \beta^{-1}\alpha x_0 k = \beta^{-1}x_0 k = \beta^{-1}\beta x_0 = x_0$ and $\beta^{-1}\alpha\beta \in A$; i.e., A is normal. We let κ be as in (1) and let $k \in K$ then $\phi(x_0 k) = y_0$. Hence by I.4.1.(3) there exists $\beta \in F$ such that $x_0 k = \beta x_0 = x_0\kappa(\beta)$ and it follows that as automorphisms of X $\kappa(\beta) = k$. Thus κ is a homomorphism of F onto K with kernel A.

Next we show that κ is τ-continuous. Let C be a closed subset of K and let $D = \{\alpha \in F \mid \kappa(\alpha) \in C\}$. If $\alpha \in (u \circ D) \cap G = \bar{D}^{\tau}$ then there are nets $\{t_i\}$ in T and $\{\alpha_i\}$ in D such that $\lim t_i = u$ and $\lim t_i\alpha_i = \alpha$. Now we can assume that $\lim \kappa(\alpha_i) = k \in C$ exists and then

$$\alpha x_o = \lim t_i\alpha_i x_o = \lim t_i x_o \kappa(\alpha_i) =$$

$$= ux_o k = x_o k,$$

by the joint continuity. Thus $\kappa(\alpha) = k \in C$ and $\alpha \in D$, so that D is τ-closed. Now $\tilde{\kappa}: F/A \to K$ is a continuous isomorphism and since K is compact Hausdorff it follows that $\tilde{\kappa}$ is a topological isomorphism. In particular F/A is Hausdorff and by 1.9.(3). $A \supset H(F)$.

(3) This follows from (2) and I.4.2.

(4) Suppose $A \supset H(F)$. Let $(Z,z_o) = \vee\{(X,\alpha x_o) \mid \alpha \in F\}$ (See Chapter I § 4). Then $\mathfrak{G}(X,\alpha x_o) = \alpha A\alpha^{-1}$ and hence $\mathfrak{G}(Z,z_o) = \cap\{\alpha A\alpha^{-1} | \alpha\in F\} = \tilde{A}$. Now clearly $\tilde{A}$ is normal in F and $\tilde{A} \supset H(F)$; by (2) $(Z,z_o) \to (Y,y_o)$, which is distal, is a group extension and therefore ϕ is almost periodic.

Conversely suppose ϕ is almost periodic then there exists a group extension $(T,Z,z_o,K) \xrightarrow{\chi} (Y,y_o)$ and a homomorphism $(Z,z_o) \xrightarrow{\psi} (X,x_o)$ such that $\chi = \phi \circ \psi$. Now let $\mathfrak{G}(Z,z_o) = B$ then by (2) $A \supset B \supset H(F)$.

//

We denote $H(G)$ by H.

IX.3. THE QUASIFACTORS $\mathfrak{Q}(F)$ OF M

Let (X,x_o) be a pointed minimal flow and let F be its Ellis group. Then since $F \subset u(u \circ F) \subset F$, F is τ-closed. Conversely let F be a τ-closed subgroup of G and set

$$\mathfrak{A}(F) = \{p \circ F \mid p \in M\} \subset 2^M.$$

Then we can prove the following.

<u>3.1. PROPOSITION</u>: <u>The flow</u> $(T, \mathfrak{A}(F))$ <u>is minimal and if we choose</u> $u \circ F$ <u>to be the base point then</u> $\mathfrak{J}(\mathfrak{A}(F), u \circ F) = F$.

<u>Proof</u>: Clearly $(T, \mathfrak{A}(F))$ is minimal. Now $\mathfrak{J}(\mathfrak{A}(F), u \circ F) = \{\alpha \in G \mid \alpha \circ F = u \circ F\}$, and $\alpha \circ F = u \circ F$ implies $\alpha \in (u \circ F) \cap G = \mathrm{cls}_\tau F = F$. Conversely if $\alpha \in F$ then by 1.12.(1) we have $\alpha \circ F = u \circ F$ and $F \subset \mathfrak{J}(\mathfrak{A}(F), u \circ F)$. //

<u>3.2. COROLLARY</u>: <u>A subgroup</u> F <u>of</u> G <u>is</u> τ-<u>closed iff</u> $F = \mathfrak{J}(X, x_o)$ <u>for some minimal flow</u> X.

<u>3.3. PROPOSITION</u>: <u>Let</u> F <u>be a</u> τ-<u>closed subgroup of</u> G <u>and let</u> ϕ: $(M,u) \to (\mathfrak{A}(F), u \circ F)$ <u>denote the homomorphism</u>: $\phi(p) = p \circ (u \circ F) = p \circ F$. <u>Then</u>

(1) $\phi^{-1}(\{p \circ F\}) = p \circ F$ <u>for every</u> $p \in M$.

(2) <u>If</u> (Y, y_o) <u>is a minimal flow such that</u> $\mathfrak{J}(Y, y_o) = F$ <u>and</u> $p \xrightarrow{\psi} py_o$ <u>is the homomorphism of</u> (M,u) <u>onto</u> (Y, y_o) <u>then there exists a proximal homomorphism</u> θ <u>from</u> $\mathfrak{A}(F)$ <u>onto</u> Y <u>such that</u> $\psi = \theta \circ \phi$. (i.e. $(\mathfrak{A}(F), u \circ F)$ <u>is the universal proximal extension of minimal flows with Ellis group</u> F.) <u>In particular</u> $\mathfrak{A}(G)$ <u>is isomorphic to the universal minimal proximal flow</u> $\Pi(T)$.

<u>Proof</u>: (1) This is obvious from 1.12.

(2) If $\phi(p) = \phi(q)$, then $p \circ F = q \circ F$ and in particular $q \in p \circ F$. Hence there exist nets $\{t_i\}$ in T and $\{\alpha_i\}$ in F such that $\lim t_i = p$ and $q = \lim t_i \alpha_i$. Thus $\psi(q) = qy_o = \lim t_i \alpha_i y_o =$

$\lim t_i y_o = py_o = \psi(p)$, and we can define $\theta(p \circ F) = py_o$ unambiguously. Clearly θ is continuous and commutes with T. Finally θ is proximal by proposition I.4.1.(2). //

3.4. LEMMA: *Let F be a τ-closed subgroup of G. If $q \in p \circ F$, then $up^{-1}q \in F$ and for every relative τ-neighbourhood N of $up^{-1}q$ in F, $q \in p \circ N$.*

Proof: Since F is τ-closed $q \in p \circ F$ implies $up^{-1}q \in (u \circ F) \cap G = F$. By lemma 2.1.(1) there are nets $\{t_i\}$ in T and $\{\alpha_i\}$ in F such that $p = \lim t_i$ and $q = \lim t_i \alpha_i$. It suffices to show that $\{\alpha_i\}$ has a subnet τ-convergent to $up^{-1}q$. Let $A_j = \{\alpha_i \mid i \geq j\}$. Clearly $q \in p \circ A_j$ which implies that $up^{-1}q \in (u \circ A_j) \cap G = \overline{A}_j^{\tau}$. This is equivalent to the existence of a subnet τ-convergent to $up^{-1}q$. //

IX.4. G/H IS THE GENERALIZED BOHR COMPACTIFICATION OF THE DISCRETE GROUP T.

We now use the results of the last three sections to describe the universal compactification flow and the generalized Bohr compactification of T.

We define the following relation on M:

$$p \sim q \quad \text{iff} \quad p \circ G = q \circ G \quad \text{and} \quad up^{-1}q \in H.$$

4.1. PROPOSITION: *The relation $\sim$ is a closed T-invariant equivalence relation on M.*

Proof: Clearly $\sim$ is an equivalence relation. Suppose $p \sim q$ and $t \in T$. Then $tp \circ G = tq \circ G$. Since $q \in p \circ G$, by lemma 3.4., $q \in p \circ N$ and $tq \in tp \circ N$ where N is any τ-neighbourhood of $up^{-1}q$ in G. This implies $u(tp)^{-1}tq \in u \circ N$ and $u(tp)^{-1}tq \in \overline{N}^{\tau}$. In particular this is true for every τ-neighbourhood N of H, because $up^{-1}q \in H$. Now, since G/H is a Hausdorff space $H = \cap \overline{N}^{\tau}$ where N runs over all the τ-neighbourhoods of H in G, and we conclude that $u(tp)^{-1}tq \in H$; i.e., $tp \sim tq$.

Next suppose that $p_i \sim q_i$ where $\{p_i\}$ and $\{q_i\}$ are nets in M such that $\lim p_i = p$ and $\lim q_i = q$. Clearly $p_i \circ G = q_i \circ G$ implies $p \circ G = q \circ G$. Let N be a τ-neighbourhood of H, then by lemma 3.4. $p_i \circ G = q_i \circ G$ implies $q_i \in p_i \circ N$, (because $up_i^{-1}q_i \in H \subset N$). Hence $q \in p \circ N$ and $up^{-1}q \in (u \circ N) \cap G = \overline{N}^{\tau}$. As above we conclude that $up^{-1}q \in H$ and $p \sim q$. Thus $\sim$ is also closed and the proof is completed.

//

Let $Z = M/\sim$ then by proposition 4.1. (T,Z) is a minimal flow. We denote by $\hat{p}$ the canonical image of p in Z.

4.2. THEOREM: There exists a jointly continuous free action of the compact Hausdorff topological group G/H on Z which commutes with the action of T on Z. The flow $(T,Z/(G/H))$ is the universal minimal proximal flow and the extension $(T,Z,G/H) \to (T,\pi(T))$ is the universal group extension of $\pi(T)$. In other words Z is isomorphic to $\pi^{\#}$ the universal compactification flow for T and G/H is the generalized Bohr compactification of T. Finally $\mathfrak{g}(\dot{Z},\hat{u}) = H$.

Proof: Clearly the map $\hat{p} \xrightarrow{\phi} p \circ G$ is a homomorphism of $(Z,\hat{u})$ onto $(\mathfrak{M}(G), u \circ G)$ which we know is the universal minimal proximal flow (3.3.(2).).

Now $\mathfrak{g}(\mathfrak{M}(G), u \circ G) = G$ and $\mathfrak{g}(Z,\hat{u}) = \{\alpha \in G \mid \alpha\hat{u} = \hat{u}\} = \{\alpha \in G \mid \hat{\alpha} = \hat{u}\}$, but $\hat{\alpha} = \hat{u}$ iff $\alpha \in H$ and $\alpha \circ G = u \circ G$. Since the latter is always true for $\alpha \in G$ $\mathfrak{g}(Z,\hat{u}) = H$. Thus to complete the proof of the theorem it suffices, by theorem 2.1., to show that ϕ is distal. So let $\hat{p}$ and $\hat{q}$ be elements of Z such that $\phi(\hat{p}) = \phi(\hat{q})$ i.e. $p \circ G = q \circ G$. This implies that $\alpha = up^{-1}q \in G$. Let $v \in J$ be such that $vp = p$ (then $pp^{-1} = v$). We will show that $v\hat{q} = \hat{q}$. By proposition I.3.2.(3). it will then follow that $\hat{q}$ and $\hat{p}$ are distal points of Z. Since $\hat{vq} = v\hat{q}$ we have to show that $\widehat{vq} = \hat{q}$. Indeed $p\alpha = pup^{-1}q = vq$ hence

$$vq \circ G = p\alpha \circ G = p \circ \alpha \circ G = p \circ G = q \circ G$$

and

$$uq^{-1}(vq) = u \in H.$$

Thus $\widehat{vq} = \hat{q}$ and the proof is completed. //

4.3. COROLLARY: *Let* T *be a strongly amenable group* (i.e. $\pi(T)$ *is trivial*); *then* $\Sigma(T) = \Gamma(T)$, *the generalized Bohr compactification of* T *coincides with its Bohr compactification. Thus the Ellis group of the universal equicontinuous minimal flow of* T *is* H.

We remark that at least in the category of topological groups, the converse is not true. There is an example in [12] of a topological (non-discrete) group T which is not strongly amenable and yet $\Gamma(T) = \Sigma(T)$.

CHAPTER X

PI-FLOWS

In this chapter we canonically associate, with each minimal flow X minimal flows X_∞ and Y_∞ such that (i) X_∞ is a proximal extension of X, (ii) X_∞ is a RIC extension of Y_∞ with the property that the only almost periodic extension of Y_∞ which is a factor of X_∞ is the trivial one and (iii) Y_∞ is a strictly PI-flow. Moreover X is a PI-flow iff $X_\infty = Y_\infty$. Theorem 2.1 (which is a "relative version of proposition IX.4.1) and proposition 3.2. constitute the building blocks for this construction.

In sections 5 and 6 new tools are developed which will enable us to show that under various conditions on the flow X the map $X_\infty \xrightarrow{\phi_\infty} Y_\infty$ is necessarily an isomorphism; i.e. that X is PI. This is done in section 7. The sources for this chapter are [12] and recent work of the author.

X.1. RIC-EXTENSIONS.

By proposition I.4.1.(3) an extension $(X,x_o) \xrightarrow{\phi} (Y,y_o)$ with $\mathcal{G}(Y,y_o) = F$ is distal iff for every $p \in M$

$$\phi^{-1}(py_o) = pFx_o.$$

In particular a minimal flow (X,x_o) is distal iff $Gx_o = X$. We say that a minimal flow (X,x_o) is incontractible if $u \circ Gx_o = X$. We say that an extension $(X,x_o) \xrightarrow{\phi} (Y,y_o)$ is <u>relatively incontractible</u> (RIC) if for every $p \in M$

$$\phi^{-1}(py_o) = p \circ Fx_o$$

where $F = \mathcal{G}(Y,y_o)$. Note that $p \circ FX_o$ is not ambiguous since an easy calculation shows that $(p \circ F)x_o = p \circ (Fx_o)$.

1.1. LEMMA: Let $(X,x_o) \xrightarrow{\phi} (Y,y_o)$ be a RIC-extension then ϕ is open and for every $p \in M$

$$\phi^{-1}(py_o) = p\circ Fx_o = p\circ\phi^{-1}(y_o).$$

Proof: The map ϕ is open iff the map $y \to \phi^{-1}(y)$ of Y into 2^X is continuous. Now $\phi^{-1}(y_o) = u\circ Fx_o$ hence

$$\phi^{-1}(py_o) = p\circ Fx_o = p\circ u\circ Fx_o = p\circ\phi^{-1}(y_o)$$

and it follows that $y \to \phi^{-1}(y)$ is continuous.//

1.2. LEMMA: A distal extension is RIC.

Proof: Let $(X,x_o) \xrightarrow{\phi} (Y,y_o)$ be distal and $p \in M$, then by I.4.1.(3).

$$\phi^{-1}(py_o) = pFx_o \subseteq (p\circ F)x_o = p\circ Fx_o = p\circ\phi^{-1}(y_o) \subseteq \phi^{-1}(py_o).//$$

Another example of a RIC-extension is the map $p \to p\circ F$ of M onto $\mathfrak{A}(F)$, indeed this is exactly the statement of proposition 3.3.(1) of chapter IX. Moreover if one examines the proof of proposition IX.4.1. it becomes clear that the proof depends upon the fact that the map $p \to p\circ G$ of M onto $\mathfrak{A}(G)$ is RIC. We shall imitate this argument in the proof of theorem 2.1 in the next section. However before we do that we pause to prove the following characterization of RIC-extensions.

Let $(X,x_o) \xrightarrow{\phi} (Y,y_o)$ and $(Z,z_o) \xrightarrow{\psi} (Y,y_o)$ be two homomorphisms of pointed minimal flows. We say that X and Z are disjoint over Y if

$$R = \{(x,z) \mid \phi(x) = \psi(z)\}$$

is a minimal subset of $X \times Z$.

1.3. PROPOSITION: The homomorphism $(X,x_o) \xrightarrow{\phi} (Y,y_o)$ is RIC iff X and $\mathfrak{A}(F)$ are disjoint over Y, where $F = \mathcal{G}(Y,y_o)$. In particular a *minimal flow* X *is incontractible iff it is disjoint from every minimal proximal flow.*

Proof: By IX.3.3.(3) there exists a homomorphism $\theta\colon (\mathfrak{A}(F),u\circ F) \to (Y,y_o)$, $\theta(p\circ F) = py_o$. Let

$$R \;=\; \{(x,p\circ F) \mid \phi(x) \;=\; \theta(p\circ F) \;=\; py_o\},$$

and denote by $\psi\colon M \to \mathfrak{A}(F)$ the homomorphism $\psi(p) = p\circ F$. Since ψ is RIC (IX.3.3.(1)) $\psi^{-1}(\{p\circ F\}) = p\circ F$ for every $p \in M$. Now

$$\Delta \;=\; \overline{\mathcal{O}}(x_o,u\circ F) \;=\; M(x_o,u\circ F \;=\; \{(px_o,p\circ F) \mid p \in M\}$$

is a minimal set and clearly $\Delta \subseteq R$. If for a subset W of a product space $V\times U$ we put $W[u] = \{v \mid (v,u) \in W\}$ then for every $y = py_o \in Y$

$$\phi^{-1}(py_o) \;=\; R[p\circ F] \quad \text{and}$$

$$\Delta[p\circ F] \;=\; \{qx_o \mid q\circ F \;=\; p\circ F\} \;=\; \psi^{-1}(\{p\circ F\})x_o$$

$$\;=\; (p\circ F)x_o \;=\; p\circ Fx_o.$$

Thus ϕ is RIC iff $\Delta = R$.//

X.2. STILL MORE ABOUT ALMOST PERIODIC EXTENSIONS.

2.1. THEOREM: Let $(X,x_o) \xrightarrow{\phi} (Y,y_o)$ *be a* RIC-*extension. Then there exists an almost periodic extension* $(Z,z_o) \xrightarrow{\psi} (Y,y_o)$ *such that* $\mathcal{G}(Z,z_o) = H(F)A$, *where* $F = \mathcal{G}(Y,y_o)$ *and* $A = \mathcal{G}(X,x_o)$, *and a homomorphism* $(X,x_o) \xrightarrow{\lambda} (Z,z_o)$. (Z,z_o) *is universal in the sense that whenever*

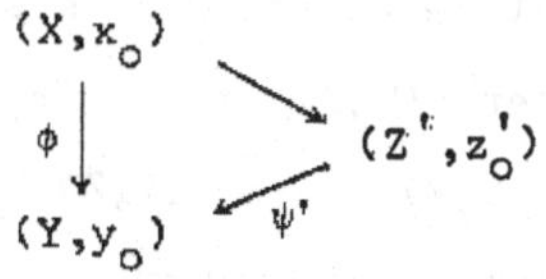

is a commutative diagram with ψ' *almost periodic then there exists a homomorphism* $(Z,z_o) \xrightarrow{\eta} (Z',z_o')$. *The extension* ψ *is an isomorphism iff* $H(F)A = F$.

Proof: Since $A \subseteq F$ and since $H(F)$ is normal in F, $H(F)A$ is a subgroup of F. By IX.1.10. $H(F)A$ is τ-closed. Define a relation on X: $x_1 \sim x_2$ iff $\phi(x_1) = \phi(x_2)$ and $up^{-1}q \in H(F)A$ for some $p,q \in M$ such that $px_o = x_1$ and $qx_o = x_2$. If also $rx_o = x_1$ and $sx_o = x_2$ for $r,s \in M$ then $up^{-1}r$ and $uq^{-1}s$ are in A and hence

$$(up^{-1}r)^{-1}(up^{-1}q)(uq^{-1}s) = ur^{-1}s \in H(F)A.$$

Thus $\sim$ is well defined and clearly it is an equivalence relation on X. We next show that it is T-invariant and closed. Let $x_1 \sim x_2$, $x_1 = px_o$ $x_2 = qx_o$ and let $t \in T$. Then $\phi(x_1) = \phi(x_2)$ implies $\phi(tx_1) = \phi(tx_2)$ and since ϕ is RIC $\phi(x_1) = \phi(x_2)$ implies

$$qx_o = x_2 \in \phi^{-1}(\phi(x_1)) = \phi^{-1}(py_o) = p\circ\phi^{-1}(y_o) = p\circ Fx_o .$$

Thus there exists $r \in p\circ F$ such that $rx_o = qx_o$, and we can assume that $q \in p\circ F$. By lemma IX.3.4. $q \in p\circ N$ and $tq \in tp\circ N$ when N is an arbitrary τ-neighbourhood of $up^{-1}q$ in F. It follows that $u(tp)^{-1}tq \in u\circ N$ and $u(tp)^{-1}tq \in \overline{N}^{\tau}$. Now $up^{-1}q \in H(F)A$ and therefore every τ-neighbourhood N of $H(F)A$ in F is also a τ-neighbourhood of $up^{-1}q$ in F. Since $F/H(F)A$ is a Hausdorff space $H(F)A = \cap \overline{N}^{\tau}$ where N runs over all the τ-neighbourhoods of $H(F)A$ in F. Thus

$$u(tp)^{-1}tq \in \cap \overline{N}^{\tau} = H(F)A$$

and $tx_1 \sim tx_2$. The fact that $\sim$ is closed is proved similarly.

Now let Z be $X/\sim$, then (T,Z) is a minimal flow. We denote the natural homomorphism of (T,X) onto (T,Z) by λ and $\lambda(x_o) = z_o$. We define $(Z,z_o) \xrightarrow{\psi} (Y,y_o)$ by $\psi(pz_o) = py_o$; clearly ψ is well defined and we have

$$\mathcal{G}(Z,z_o) = \{\alpha \in G \mid \alpha z_o = z_o\} = \{\alpha \in G \mid \lambda(\alpha x_o) = \lambda(x_o)\}$$

$$= \{\alpha \in G \mid \phi(\alpha x_o) = \alpha y_o = y_o = \phi(x_o) \text{ and } \alpha \in H(F)A\} =$$

$$= \{\alpha \in G \quad \alpha \in H(F)A\} = H(F)A.$$

In order to show that ψ is almost periodic it suffices, by theorem IX.2.1.(4), to show that ψ is distal. Let $z_1, z_2 \in Z$ and $\psi(z_1) = \psi(z_2)$, suppose $z_1 = pz_o$ and $z_2 = qz_o$ then $py_o = \psi(pz_o) = \psi(qz_o) = qy_o$ and $up^{-1}q = \alpha \in F$. Let $v \in J$ be such that $vp = p$, we will show that $vz_2 = z_2$ and since also $vz_1 = vpz_o = pz_o = z_1$ it will follow from I.3.2.(3) that z_1 and z_2 are distal. Now let $x_2 = qx_o$ then $\lambda(vx_2) = \lambda(x_2)$ indeed $p\alpha = pup^{-1}q = vq$ hence $vqy_o = p\alpha y_o = py_o = qy_o$ and $uq^{-1}(vq) = u \in H(F)A$ i.e. $vx_2 = vqx_o \sim qx_o = x_2$ or $\lambda(vx_2) = \lambda(x_2)$. Therefore $vz_2 = v\lambda(x_2) = \lambda(vx_2) = \lambda(x_2) = z_2$. This completes the proof of the existence and almost periodicity of ψ. The universality now follows from IX.2.1.(4), I.4.1.(1), and I.4.2.//

X.3. A CONSTRUCTION OF RIC-EXTENSIONS

Let $(X,x_o) \xrightarrow{\phi} (Y,y_o)$ be a homomorphism of minimal flows with $\mathcal{G}(Y,y_o) = F$. We define a quasifactor of X,

$$Y' = \{p \circ Fx_o \mid p \in M\}, \qquad y_o' = u \circ Fx_o.$$

Also let $(X',x_o') = (X,x_o) \vee (Y',y_o')$, thus $x_o' = (x_o,y_o')$. Clearly Y' and X' depends only on F and X, and we shall write $Y' = \mathfrak{A}(X,F)$, $X' = \mathcal{L}(X,F)$.

<u>3.1. LEMMA</u>: <u>Let</u>

$$\Delta = \{(x,y') \mid x \in X', \; y' \in Y' \text{ and } x \in y'\}$$

<u>Then</u> $X' = \Delta$.

<u>Proof</u>: Clearly $x_o' = (x_o, u \circ Fx_o)$ is in Δ which is a closed

invariant subset of $X \times Y'$. Hence $\overline{0}(x_o') = X' \subseteq \Delta$. We complete the proof by showing that also $\Delta \subseteq X'$. Now

$$Fx_o' = F(x_o, y_o') = F(x_o, u \circ Fx_o) = \{(\alpha x_o, \alpha \circ Fx_o) \mid \alpha \in F\}$$

$$= \{\alpha x_o \mid \alpha \in F\} \times \{u \circ Fx_o\} = \{\alpha x_o \mid \alpha \in F\} \times \{y_o'\}.$$

Hence

$$p \circ Fx_o' = p \circ Fx_o \times \{py_o'\} = p \circ Fx_o \times \{p \circ Fx_o\}$$

$$= \{(x, py_o') \mid x \in py_o'\}$$

Therefore

$$\Delta = \{(x,y') \mid x \in y'\} = \bigcup_{p \in M} \{(x, py_o') \mid x \in py_o'\} = \bigcup_{p \in M} p \circ Fx_o'.$$

Clearly the last set is contained in the orbit closure of x_o' i.e. $\Delta \subseteq X'$. //

Define maps:

$$\theta : Y' \to Y \qquad \theta(\{p \circ Fx_o\}) = py_o$$

$$\theta' : X' \to X \qquad \theta'((x,y')) = x$$

$$\text{and} \quad \phi' : X' \to Y' \qquad \phi'((x,y')) = y'.$$

<u>3.2 PROPOSITION</u>: (1) θ <u>is a well defined proximal homomorphism.</u>

(2) <u>The diagram</u>

$$\begin{array}{ccc} X' & \xrightarrow{\theta'} & X \\ \phi' \downarrow & & \downarrow \phi \\ Y' & \xrightarrow{\theta} & Y \end{array}$$

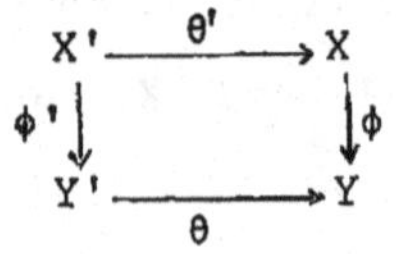

<u>is commutative</u>, θ' <u>is proximal</u>, ϕ' <u>is</u> RIC, <u>and</u> θ <u>is an isomorphism iff</u> ϕ <u>is</u> RIC.

<u>Proof</u>: (1) Suppose $p, q \in M$ and $p \circ Fx_o = q \circ Fx_o$, since $Fx_o \subseteq \phi^{-1}(y_o)$ it follows tha

$$p \circ Fx_o \subseteq p \circ \phi^{-1}(y_o) \subseteq \phi^{-1}(py_o)$$

and

$$q \circ Fx_o \subseteq q \circ \phi^{-1}(y_o) \subseteq \phi^{-1}(qy_o).$$

Thus $\phi^{-1}(py_o) \cap \phi^{-1}qy_o) \neq \emptyset$, $py_o = qy_o$ and θ is well defined. If $\theta(p\circ Fx_o) = \theta(q\circ Fx_o)$ i.e. $py_o = qy_o$ then $up^{-1}qy_o = y_o$ and $\alpha = up^{-1}q \in F$. Now $up^{-1}\circ(p\circ Fx_o) = u\circ Fx_o$ and by IX.1.12.

$$up^{-1}\circ(q\circ Fx_o) = \alpha\circ Fx_o = (\alpha\circ F)x_o = u\circ Fx_o.$$

It follows that $p\circ Fx_o$ and $q\circ Fx_o$ are proximal points of Y' and θ is proximal.

(2) The maps θ' and ϕ' are the restrictions to X' of the projection maps of $X \times Y'$ onto X and Y' respectively hence they are both homomorphisms and the fact that the diagram is comutative follows because the flows are pointed.

Let $x', x_1' \in X'$ and suppose $x' = p(x_o,y_o')$ and $x_1' = q(x_o,y_o')$; if $px_o = \theta'(x') = \theta'(x_1') = qx_o$ then $py_o = qy_o$ and hence $\theta(py_o') = \theta(qy_o')$. Since θ is proximal if follows that py_o' and qy_o' are proximal and therefore x' and x_1' which have the same first coordinate are proximal. Next we show that ϕ' is RIC. Let $p \in M$, then by lemma 3.1.

$$\begin{aligned}
\phi'^{-1}(py_o') &= \{(x,py_o') \mid (x,py_o') \in X'\} = \\
&= \{(x,py_o') \mid (x,py_o') \in \Delta\} = \\
&= \{(x,py_o') \mid x \in py_o'\} = \\
&= \{(x,py_o') \mid x \in p\circ Fx_o\} = \\
&= (p\circ Fx_o) \times \{py_o'\} = \\
&= p\circ((Fx_o) \times \{y_o'\}) = p\circ F(x_o,y_o') = p\circ Fx_o'.
\end{aligned}$$

Thus ϕ' is RIC. It now follows that if θ is an isomorphism ϕ is RIC, and conversely if ϕ is RIC, $\phi^{-1} : Y \to Y' \subseteq 2^X$ is continuous (1.1.) and since $\phi^{-1}(py_o) = p\circ Fx_o$, ϕ^{-1} is a homomorphism of Y onto Y'. Clearly $\theta\circ\phi^{-1}$ is the identity map on Y and hence θ is an isomorphism. //

X.4. PI-FLOWS

We say that a minimal flow X is <u>strictly</u> PI if there is an ordinal ν and flows $\{(W_\alpha, w_\alpha)\}_{\alpha \le \nu}$ such that (i) W_0 is the trivial flow. (ii) For every $\alpha < \nu$ there exists a homomorphism $(W_{\alpha+1}, w_{\alpha+1}) \xrightarrow{\phi_\alpha} (W_\alpha, w_\alpha)$ which is either proximal or almost periodic. (iii) For a limit ordinal $\alpha \le \nu$ $(W_\alpha, w_\alpha) = \bigvee_{\beta<\alpha}(W_\beta, w_\beta)$. (iv) $W_\nu = X$.

We say that X is a PI-<u>flow</u> if there exist a strictly PI flow X' and a proximal homomorphism $X' \xrightarrow{\phi} X$. It follows from the Furstenberg structure theorem for distal flows [13] that every metric distal flow is strictly PI and from the Veech-Ellis theorem ([41], [10]) it follows that every metric point-distal flow is PI. (A minimal flow X is <u>point-distal</u> if there exists a point $x_0 \in X$ which is proximal only to itself.)

Next we associate with each pointed minimal flow (X, x_0) a canonical construction of a proximal extension $X_\infty \to X$ and a strictly PI-flow Y_∞ which is a maximal PI factor of X_∞. To this end we need the following lemma.

<u>4.1 LEMMA</u>: <u>Let</u> A <u>and</u> B <u>be</u> τ-<u>closed subgroups of</u> G. <u>If</u> B <u>is a normal subgroup, then</u> BA <u>is a</u> τ-<u>closed subgroup and</u> $H(BA)A = H(B)A$.

<u>Proof</u>: By IX.1.10., BA is τ-closed, and it is a subgroup because B is normal.

From the definition of the functor H it is clear that $H(BA) \supseteq H(B)$, hence $H(BA)A \supseteq H(B)A$. We shall show that $BA/H(B)A$ with the quotient τ-topology is Hausdorff. This will imply, by IX.1.9.3., that $H(B)A \supseteq H(BA)$ whence $H(B)A = H(BA)A$.

After noting that $H(B)$ is also a τ-closed normal subgroup of G, consider the map $B/(B \cap H(B)A) \xrightarrow{\phi} BA/H(B)A$ defined by $\phi(\beta(B \cap H(B)A)) = \beta H(B)A$, for $\beta \in B$. Clearly ϕ is well defined one-to-one and τ-continuous. If C is a closed subset of $B/(B \cap H(B)A)$ and $\tilde{C}$ is its counter image in B then $\tilde{C}$ is τ-closed and by IX.1.10. $\tilde{C}H(B)A$ is τ-closed, hence $\phi(C)$ is closed in $BA/H(B)A$. Thus ϕ is also a closed map hence a τ-homeomorphism. Since $H(B) \subseteq B \cap H(B)A$ it follows that $B/(B \cap H(B)A)$ is Hausdorff and therefore so is $BA/H(B)A$. The proof is complete.//

We now define inductively the following decreasing sequence of τ-closed normal subgroups of G. Put $G_o = G$, $G_1 = H(G) = H$ and for every ordinal α let $G_{\alpha+1} = H(G_\alpha)$. If α is a limit ordinal put $G_\alpha = \bigcap_{\beta<\alpha} G_\beta$. By IX.1.9.(1), $H(F)$ is invariant under every topological automorphisms of F, where F is any τ-closed subgroup of G. Inductively one shows that each G_α is invariant under the inner-automorphisms of G. In other words each G_α is a normal subgroup of G.

Next we define inductively the canonical PI-tower associated with (X,x_o). Put $A = \mathfrak{G}(X,x_o)$ and let $X_o = X$ and Z_o be the trivial one point flow. We let

$$\begin{array}{rcccc} \mathcal{L}(X,G) & = & (X_1,x_1) & \xrightarrow{\psi_o'} & (X_o,x_o) \\ & & \phi_o' \downarrow & & \downarrow \phi_o \\ \mathfrak{A}(X,G) & = & (Y_1,y_1) & \xrightarrow[\psi_o]{} & Z_o \end{array}$$

be the diagram constructed in section 3. Thus ψ_o and ψ_o' are proximal extension and hence $\mathfrak{G}(X_1,x_1) = \mathfrak{G}(X_o,x_o) = A$, $\mathfrak{G}(Y_1,y_1) = \mathfrak{G}(Z_o) = G$ and ϕ_o' is RIC. By theorem 2.1. there exists a diagram

$$\begin{array}{ccc} & & (X_1,x_1) \\ & \phi_1 \swarrow & \downarrow \phi_o' \\ (Z_1,z_1) & \xrightarrow[\chi_1]{} & (Y_1,y_1) \end{array}$$

where χ_1 is almost periodic and $\mathfrak{G}(Z_1,z_1) = H(G)A = G_1A$.

Let β be an ordinal and suppose that we have already constructed flows (X_β, x_β) and (Z_β, z_β) such that $\mathcal{G}(X_\beta, x_\beta) = A$, $\mathcal{G}(Z_\beta, z_\beta) = G_\beta A$ (Recall that G_β is normal so that $G_\beta A$ is a τ-closed subgroup) and a homomorphism $(X_\beta, x_\beta) \xrightarrow{\phi_\beta} (Z_\beta, z_\beta)$. We construct a diagram

$$\begin{array}{ccccc}
\mathcal{L}(X_\beta, G_\beta A) & = & (X_{\beta+1}, x_{\beta+1}) & \xrightarrow{\psi'_\beta} & (X_\beta, x_\beta) \\
 & & \phi'_\beta \downarrow & & \downarrow \phi'_\beta \\
\mathcal{A}(X_\beta, G_\beta A) & = & (Y_{\beta+1}, y_{\beta+1}) & \xrightarrow[\psi_\beta]{} & (Z_\beta, z_\beta)
\end{array}$$

as in section 3. Thus ψ_β and ψ'_β are proximal, $\mathcal{G}(X_{\beta+1}, x_{\beta+1}) = \mathcal{G}(X_\beta, x_\beta) = A$, $\mathcal{G}(Y_{\beta+1}, y_{\beta+1}) = \mathcal{G}(Z_\beta, z_\beta) = G_\beta A$ and ϕ'_β is RIC. By theorem 2.1. there exists a diagram

$$\begin{array}{ccc}
 & \overset{\phi_{\beta+1}}{\swarrow} & (X_{\beta+1}, x_{\beta+1}) \\
 & & \downarrow \phi'_\beta \\
(Z_{\beta+1}, z_{\beta+1}) & \xrightarrow[\chi_{\beta+1}]{} & (Y_{\beta+1}, y_{\beta+1})
\end{array}$$

where $x_{\beta+1}$ is almost periodic and $\mathcal{G}(Z_{\beta+1}, z_{\beta+1}) = H(G_\beta A)A$. By lemma 4.1. $H(G_\beta A)A = H(G_\beta)A = G_{\beta+1}A$.

If α is a limit ordinal we define $(Z_\alpha, z_\alpha) = \bigvee_{\beta<\alpha} (Z_\beta, z_\beta)$ and $(X_\alpha, x_\alpha) = \bigvee_{\beta<\alpha} (X_\beta, x_\beta)$. Then

$$\mathcal{G}(X_\alpha, x_\alpha) = \bigcap_{\beta<\alpha} \mathcal{G}(X_\beta, x_\beta) = A \qquad \text{(I.4.3.)}$$

and by IX.1.11.

$$\mathcal{G}(Z_\alpha, z_\alpha) = \bigcap_{\beta<\alpha} \mathcal{G}(Z_\beta, z_\beta) = \bigcap_{\beta<\alpha} (G_\beta A) =$$

$$= (\bigcap_{\beta<\alpha} G_\beta)A = G_\alpha A.$$

Finally we let $(X_\alpha, x_\alpha) \xrightarrow{\phi_\alpha} (Z_\alpha, z_\alpha)$ be the natural map. The construction is carried on inductively. Notice that Y_β is defined only for non-limit ordinals. There exists a least non-limit ordinal $\eta = \eta(X, x_0)$ for which the extension $Z_\eta \xrightarrow{\chi_\eta} Y_\eta$ is an isomorphism. Then $X_\eta \xrightarrow{\phi_\eta} Z_\eta$

is RIC and by 3.2.(2) ψ_η and ψ'_η are also isomorphisms. Thus for every $\delta \geq \eta$ all the maps ψ_δ, ψ_δ and χ_δ in the tower will be isomorphisms. Now by theorem 2.1. χ_η is an isomorphism iff

$$G_\eta A = \mathcal{G}(Z_\eta, z_\eta) = \mathcal{G}(Y_\eta, y_\eta) = G_{\eta-1} A,$$

and if this is the case then also $G_\delta A = G_\eta A$ for every $\delta \geq \eta$. It follows that η depends only on A. The family of closed subsets

$$u \circ G_\beta A x_0 = u \circ G_\beta x_0$$

is a decending family of closed subsets of X, hence if X is metrix then there are at most countably many of them. If we examine the definition of the flow $Y_{\beta+1}$ we shall see that $Y_{\beta+1}$ is isomorphic to the orbit closure of the point

$$(u \circ G x_0, u \circ G_1 x_0, \ldots, u \circ G_\alpha x_0, \ldots, u \circ G_\beta x_0) \in (2^X)^{\beta+1}.$$

Thus if X is metric η is countable and all the flows which appear in the canonical tower are metric flows. We can also conclude that if A is a τ-closed subgroup such that there is a metric minimal flow (X,x_0) with $\mathcal{G}(X,x_0) = A$, then for every minimal flow (W,w_0) with $\mathcal{G}(W,w_0) = A$, $\eta(W,w_0) = \eta(A)$ is countable.

We put $X_\eta = X_\infty$, $Y_\eta = Y_\infty$ and we call Y_∞ the <u>canonical</u> PI-<u>flow</u> <u>associated with</u> X. <u>Note that</u> (X_∞, x_∞) <u>has the form</u>

$$(X_\infty, x_\infty) = (X,x_0) \vee (Y_\infty, y_\infty)$$

$$= \bar{O}(x_0, u \circ G_1 x_0, \ldots, u \circ G_{\eta-1} x_0) \subseteq (2^X)^\eta$$

<u>and that the map</u> $(X_\infty, x_\infty) \overset{\phi_\infty}{\rightarrow} (Y_\infty, y_\infty)$ is RIC. <u>If we denote</u> $F = G_\eta A$ <u>then</u>

$$\mathcal{G}(Y_\infty, y_\infty) = F, \qquad \mathcal{G}(X_\infty, x_\infty) = A$$

<u>and</u>

$$H(F)A = H(G_\eta A)A = G_{\eta+1} A = G_\eta A = F.$$

We now let $\iota = \eta(\{u\})$ i.e. ι is the least non-limit ordinal such that $G_\iota = G_{\iota+1}$ and we write $G_\iota = G_\infty$.

4.2. THEOREM: Let X be a minimal flow. The following conditions on X are equivalent.

(1) X is a PI-flow.

(2) There exists a point $x_0 \in X$ such that $ux_o = x_o$ and $\mathcal{G}(X,x_o) \supseteq G_\infty$.

(3) For every $x_0 \in X$ such that $ux_o = x_o$, $\mathcal{G}(X,x_o) \supseteq G_\infty$.

(4) X is a factor of $\mathcal{O}\!\mathcal{l}(G_\infty)$.

(5) In the canonical tower for X, $X_\infty = Y_\infty$.

Proof: (1) ⇒ (2) Let X be a PI-flow; then there exist a proximal extension $X' \xrightarrow{\psi} X$ and an ordinal ν and a family of pointed minimal flows $\{(W_\alpha,w_\alpha)\}_{\alpha\leq\nu}$ with homomorphisms $(W_{\alpha+1},w_{\alpha+1}) \xrightarrow{\phi_\alpha} (W_\alpha, w_\alpha)$, such that each ϕ_α is either proximal or almost periodic. Also for limit ordinal α $(W_\alpha,w_\alpha) = \bigvee_{\beta<\alpha} (W_\beta,w_\beta)$ and W_0 is the trivial flow while $(W_\nu,w_\nu) = (X',x_0')$. Let $A = \mathcal{G}(X,x_o) = \mathcal{G}(X',x_0')$ and let $A_\alpha = \mathcal{G}(W_\alpha,w_\alpha)$. Then $A_0 = G$; suppose $A_\alpha \supseteq G_\alpha$ then we consider $(W_{\alpha+1},w_{\alpha+1}) \xrightarrow{\phi_\alpha} (W_\alpha,w_\alpha)$, if ϕ_α is proximal

$$A_{\alpha+1} = A_\alpha \supseteq G_\alpha \supseteq G_{\alpha+1} \qquad \text{(I.4.1.(2))}.$$

If ϕ_α is almost periodic then by IX.2.1.(4).

$$A_{\alpha+1} = \mathcal{G}(W_{\alpha+1},w_{\alpha+1}) \supseteq H(\mathcal{G}(W_\alpha,w_\alpha)) = H(A_\alpha) \supseteq H(G_\alpha) = G_{\alpha+1}.$$

If α is a limit ordinal and $G_\beta \subseteq A_\beta$ for every $\beta < \alpha$, then

$$A = \mathcal{G}(W_\alpha,w_\alpha) = \mathcal{G}(\bigvee_{\beta<\alpha}(W_\beta,w_\beta)) = \bigcap_{\beta<\alpha} A_\beta \supseteq \bigcap_{\beta<\alpha} G_\beta = G_\alpha. \qquad \text{(I.4.3)}$$

Thus by induction $\mathcal{G}(X,x_0) = A \supseteq G_\nu \supseteq G_\infty$.

(2) ⇒ (3) If $\mathcal{G}(X,x_0) \supset G_\infty$ and $x_1 \in X$ is such that $ux_1 = x_1$ then there exists $\alpha \in G$ such that $\alpha x_0 = x_1$ and then

$$\mathcal{G}(X,x_1) = \alpha\mathcal{G}(X,x_0)\,\alpha^{-1} \supseteq \alpha G_\infty \alpha^{-1} = G_\infty.$$

(2) ⇔ (4) Let $\mathcal{G}(X,x_o) \supseteq G_\infty$ and define

$$(Z,z_o) = (X,x_o) \vee (\mathcal{O}\!\ell(G_\infty), u\circ G_\infty).$$

Then $\mathcal{G}(Z,z_o) = \mathcal{G}(X,x_o) \cap G_\infty = G_\infty$. By the universality of $\mathcal{O}\!\ell(G_\infty)$ (IX.3.3.(2)), $\mathcal{O}\!\ell(G_\infty)$ is isomorphic to Z. Since X is a factor of Z it is also a factor of $\mathcal{O}\!\ell(G_\infty)$. The converse follows from I.4.1.(1).

(2) ⇒ (5) Suppose $A = \mathcal{G}(X,x_o) \cap G_\infty$ and construct $X_\infty \xrightarrow{\phi_\infty} Y_\infty$. Then ϕ_∞ is RIC and

$$\mathcal{G}(Y_\infty,y_\infty) = G_{\eta(A)}A = G_{\eta(A)+1}A = \dots = G_\infty A = A.$$

Therefore, by the definition of a RIC-extension for every $y = py_\infty \in Y_\infty$

$$\phi^{-1}(y) = p\circ Ax_\infty = \{px_\infty\}$$

Thus ϕ is one-to-one i.e. an isomorphism.

(5) ⇒ (1) By definition Y_∞ is strictly PI, hence $X_\infty = Y_\infty$ implies that X, which admits X_∞ as a proximal extension, is PI. //

We have the following corollary which is far from being a direct consequence of the definition of a PI-flow.

4.3. COROLLARY: A factor of a PI-flow is PI.

Another corollary of this theorem is the following proposition

4.4 PROPOSITION: Let X be a non-trivial PI-flow which is incontractible. Then X admits a non-trivial almost periodic factor.

Proof: Since X is PI, $X_\infty = Y_\infty$; since X is incontractible $u\circ Gx_o = X$ and Y_1 the first stage in the canonical PI-tower of Y_∞ is trivial thus $(Z_1,z_1) \xrightarrow{X_1} (Y_1,y_1)$ is an almost periodic extension of the trivial flow i.e. Z_1 is an almost periodic flow, which is a factor of $Y_\infty = X_\infty$. Since the extension $X_\infty \to X$ is proximal Z_1 is already a factor of X. Finally if Z_1 were trivial, then the whole tower would collapse forcing Y_∞, X_∞, and X to be trivial.//

X.5. A BASIS FOR THE τ-TOPOLOGY

We recall that our group T is discrete; it is therefore true that for every subset U of T, $\overline{U}$ is an open and closed subset of βT and $\overline{U} \cap T = U$. Moreover the family

$$\{\overline{U} \mid U \subseteq T\}$$

forms a basis for the topology of βT. For subsets U and V of T we put

$$[U,V] = \{p \in M \mid \text{there exists } t \in U \text{ such that } tp \in \overline{V}\}$$

and

$$(U,V) = [U,V] \cap G.$$

<u>5.1. PROPOSITION</u>: <u>The family</u>

$$\{(U,V) \mid U,V \subseteq T \text{ and } u \in \overline{U} \cap \overline{V}\}$$

<u>forms a basis for the</u> τ-<u>topology on</u> G <u>at</u> u.

<u>Proof</u>: Let W be an open τ-neighbourhood of u in G. Let $K = G\backslash W$, then K is τ-closed. Since $u \circ K$ is a closed subset of M, $\mathcal{O} = M\backslash(u \circ K)$ is open and because $(u \circ K) \cap G = \overline{K}^{\tau} = K$, $\mathcal{O}$ is an open neighbourhood of u in M. For every open neighbourhood $\mathcal{O}_i$ of u which is contained in $\mathcal{O}$ let

$$U_i = \{t \in T \mid tu \in \mathcal{O}_i\}$$

and let V_i be a subset of T such that $u \in \overline{V}_i \cap M \subset \mathcal{O}_i$. Suppose $(U_i,V_i) \cap K \neq \emptyset$ for every i. Choose α_i in this intersection, then there exists $t_i \in U_i$ with $t_i\alpha_i \in \overline{V}_i$. Since $t_i \in U_i$ we have $\lim t_i u = u$ and thus

$$u = \lim t_i\alpha_i = \lim t_i u\alpha_i \in (u \circ K) \cap G = K.$$

This is a contradiction and we conclude that for some i, $(U_i,V_i) \cap K = \emptyset$ i.e. $(U_i, V_i) \subseteq W$.

Conversely, let (U,V) such that $u \in \overline{U} \cap \overline{V}$ be given and let $K = G \setminus (U,V)$. If $\alpha \in K$ then for every $t \in U$, $t\alpha \notin \overline{V}$. Let $\beta \in (u \circ K) \cap G$ then $\beta = \lim t_i \alpha_i$ where $\lim t_i = u$ and $\alpha_i \in K$. Eventually $t_i \in \overline{U}$ which is a neighbourhood of u and hence $t_i \in \overline{U} \cap T = U$. Therefore $t_i \alpha_i \notin \overline{V}$ and since $\overline{V}$ is open $\beta \notin \overline{V}$. We have shown that $\text{cls } K = (u \circ K) \cap G \subseteq G \setminus \overline{V}$; hence

$$u \in \overline{V} \cap G \subseteq G \setminus (\text{cls}_\tau K) = \text{int}_\tau (U,V)$$

and $\text{int}_\tau(U,V)$ is an open τ-neighbourhood of u which is contained in (U,V). This completes the proof.//

We can use this representation of the τ-topology to show that the inversion in G is a τ-homeomorphism.

<u>5.2. LEMMA</u>: <u>For</u> $U,V \subseteq T$

$$(U,V)^{-1} = (V,U).$$

<u>Proof</u>: Let $\alpha \in (U,V)$ then there exists $t \in U$ such that $t\alpha \in \overline{V}$. Hence there is a net $\{s_i\}$ in V such that $\lim s_i = t\alpha$, and therefore $\lim s_i \alpha^{-1} = t$. Eventually $s_i \alpha^{-1} \in \overline{U}$ which is a neighbourhood of t and it follows that $\alpha^{-1} \in (V,U)$. We have shown that $(U,V)^{-1} \subseteq (V,U)$ and hence $(U,V) \subseteq (V,U)^{-1} \subseteq (U,V)$.//

<u>5.3. COROLLARY</u>: <u>The map</u> $\alpha \to \alpha^{-1}$ <u>of</u> G <u>onto itself is a</u> τ-<u>homeomorphism</u>.

<u>5.4. LEMMA</u>: <u>For</u> $t \in T$ <u>and</u> $U,V \subseteq T$

$$[t^{-1}U, t^{-1}V] = [U,V].$$

<u>Proof</u>:

$$\begin{aligned}[t^{-1}U, t^{-1}V] &= \{p \in M \mid \text{ there exists } s \in t^{-1}U \text{ such that } sp \in \overline{t^{-1}V}\} \\ &= \{p \in M \mid \text{ there exists } s \in T \text{ such that } ts \in U \text{ and } tsp\ \overline{V}\} \\ &= [U,V]. \text{//}\end{aligned}$$

5.5. COROLLARY : For every $v \in J$, the family

$$\mathcal{B} = \{[U,V] \cap vG \mid u \in \overline{U} \cap \overline{V}\}$$

forms a basis for the τ-topology on vG (IX section 1.)

Proof: By proposition 5.1. the collection

$$\mathcal{A} = \{[U,V] \cap vG \mid v \cap \overline{U} \cap \overline{V}\}$$

forms a basis for the τ-topology on vG. Now, there is a net $\{t_i\}$ in T such that $\lim t_i u = v$ if $v \in \overline{U} \cap \overline{V}$ then eventually $t_i u \in \overline{U} \cap \overline{V}$ and $u \in \overline{t_i^{-1}U} \cap \overline{t_i^{-1}V}$. By 5.4.

$$[U,V] \cap vG = [t_i^{-1}U, t_i^{-1}V] \cap vG$$

and we have shown that $\mathcal{A} \subseteq \mathcal{B}$. Similarly one shows that $\mathcal{B} \subseteq \mathcal{A}$ and thus $\mathcal{B} = \mathcal{A}$.//

X.6. ELLIS' TRICK

As was shown in IX.1.5. the τ-topology on G is weaker than the topology induced by M on G. This means that if W is a τ-open subset of G then there is an open set V in M such that $V \cap G = W$. The converse of course need not be true; however Ellis has shown that under certain conditions a partial converse does hold (lemma 6.1.). This relation between the τ-topology and the induced topology is a very powerfull device and was used by Ellis in [10], by Ellis and Keynes in [11] and also in [12].

Let F be a τ-closed subgroup of G. Since the map $p \to p\alpha$ for $\alpha \in F$ is continuous on M we can consider (M,F) as a flow (the action being written on the right) of the discrete group F. Clearly $\overline{F}$ is an F-invariant subset of M and therefore there exists an F-minimal subset $K \subseteq \overline{F}$. Let $p \in K$ and write $p = w\alpha$,

$w \in J$ and $\alpha \in G$. Then since $p \in \overline{F}$, $p = \lim \alpha_i$ for some net $\{\alpha_i\}$ in F and by IX.1.5. $\tau\text{-}\lim \alpha_i = up = \alpha$. Since F is τ-closed $\alpha \in F$. Now $KF \subseteq K$ hence $w = p\alpha^{-1}$ is in K, and since K is F-minimal $\overline{wF} = K$. Thus we have shown that $\overline{F}$ <u>contains an idempotent</u> w <u>such that</u> $\overline{wF}$ <u>is an</u> F-<u>minimal set</u>. We fix such $w \in \overline{F}$.

<u>6.1. LEMMA</u>: <u>Let</u> $p \in \overline{wF}$ <u>and let</u> V <u>be an open neighbourhood of</u> p <u>in</u> $\overline{wF}$. <u>Then in</u> $\overline{wF}$

$$W = \operatorname{int}_\tau \operatorname{cls}_\tau(V \cap wF) \neq \emptyset$$

<u>Proof</u>: Let $q \in \overline{wF}$, then by the F-minimality of $\overline{wF}$ there exists $\alpha \in F$ such that $q\alpha \in V$. It follows that

$$\overline{wF} \subset VF = \cup \{V\alpha \mid \alpha \in F\}.$$

Taking a finite subcovering we have $\overline{wF} \subseteq \bigcup_{i=1}^{n} V\alpha_i$ where $\alpha_i \in F$. In particular $wF \subseteq \bigcup_{i=1}^{n} (V \cap wF)\alpha_i$ and hence also

$$wF \subseteq \bigcup_{i=1}^{n} \operatorname{cls}_\tau(V \cap wF)\alpha_i.$$

This implies that at least one of the sets $\operatorname{cls}_\tau(V \cap wF)\alpha_i$ has a non-empty τ-interior, but then so does also $\operatorname{cls}_\tau(V \cap wF)$. //

We now use the last lemma to obtain some information about the group wF when F is the group met with at the top of the PI-tower built for a flow (X,x_0) with $\mathcal{G}(X,x_0) = A$. Thus we assume that $H(F)A = F$ for some τ-closed subgroup A of F.

<u>6.2. PROPOSITION</u>: <u>Let</u> A <u>be a</u> τ-<u>closed subgroup of</u> F <u>such that</u> $H(F)A = F$. <u>Let</u> U <u>be a subset of</u> T <u>such that</u> $u \in \overline{U}$. <u>Then</u>

$$L = ([U,U] \cap wF)A = [U,U]A \cap wF$$

<u>is dense in</u> $\overline{wF}$.

<u>Proof</u>: By corollary 5.5 $[U,U] \cap wF$ is a τ-neighbourhood of W in

wF. From its definition $H(wF) \subseteq cls_\tau\ int_\tau([U,U] \cap wF)$, and since $\alpha \to w\alpha$ is a τ-homeomorphism of F onto wF, $H(wF) = wH(F)$. Thus $H(wF)A = wF$ and by IX.1.10.

$$\begin{aligned} wF &= H(wF)A \subseteq (cls_\tau\ int_\tau([U,U] \cap wF))A \\ &= cls_\tau[(int_\tau([U,U] \cap wF))A] \\ &= cls_\tau\ int_\tau\ L \subset wF. \end{aligned}$$

and it follows that $cls_\tau\ int_\tau\ L = wF$. Now let p be an arbitrary point of $\overline{wF}$ and let V be an open neighbourhood of p in $\overline{wF}$. Then by lemma 6.1.

$$int_\tau\ cls_\tau(V \cap wF) \neq \emptyset.$$

Thus

$$\begin{aligned} cls_\tau\ int_\tau\ L \cap int_\tau\ cls_\tau\ (V \cap wF) &\neq \emptyset \\ int_\tau\ L \cap int_\tau\ cls_\tau\ (V \cap wF) &\neq \emptyset \\ int_\tau\ L \cap cls_\tau\ (V \cap wF) &\neq \emptyset \\ int_\tau\ L \cap (V \cap wF) &\neq \emptyset \\ L \cap V \neq \emptyset \quad \text{and} \quad p &\in \overline{L} \end{aligned}$$

We have shown therefore that L is dense in $\overline{wF}$. //

X.7. STRUCTURE THEOREMS

In this last section we shall prove that under various conditions on its proximal relation a metric minimal flow is PI. In particular a metric minimal distal flow is PI and it is then easy to see that its canonical PI tower consists of almost periodic extensions only (Furstenberg's theorem).

7.1 PROPOSITON: Let $(X,x_0) \xrightarrow{\phi} (Y,y_0)$ be a homomorphism of minimal flows. Suppose that X is metric and that $H(F)A = F$, where

$F = \mathcal{G}(Y,y_o)$ and $A = \mathcal{G}(X,x_o)$. Let $w \in \overline{F}$ be an idempotent such that $\overline{wF}$ is an F-minimal subset of M, and let $v \in J$ be an arbitrary idempotent of M. Then

(1) $$K = \{p \in \overline{wF} \mid px_o \text{ and } vx_o \text{ are proximal}\}$$

is a residual subset of $\overline{wF}$.

(2) $$W = \{x \in \overline{wFx_o} \mid x \text{ and } vx_o \text{ are proximal}\}$$

is a residual subset of $\overline{wFx_o}$.

Proof: Let $\{N_i\}$ be a countable basis for the topology at vx_o. Since the maps $q \to qx_o$ and $q \to qvx_o$ of βT into X are continuous there exists a subset U_i of T such that $v \in \overline{U}_i$ and $\overline{U}_i x_o \subseteq N_i$ and $\overline{U}_i vx_o \subseteq N_i$. We choose a net $\{s_j\}$ in T such that $\lim s_j u = v$ and then eventually $u \in \overline{s_j^{-1}U}$. Since $[s_j^{-1}U_i, s_j^{-1}U_i] = [U_i,U_i]$ (5.4.) we can conclude by lemma 6.2. that

$$L_i = [U_i,U_i]A \cap \overline{wF}$$

is dense in $\overline{wF}$. Since $[U_i,U_i]$ is open in M, so is $[U_i,U_i]A$ and therefore L_i is an open dense subset of $\overline{wF}$ and hence $L = \cap L_i$ is a residual subset of $\overline{wF}$.

If $p \in L$ then for each i there exists an $\alpha_i \in A$ such that $p\alpha_i \in [U_i,U_i]$ i.e. there exists $t_i \in U_i$ for which $t_i p\alpha_i \in \overline{U}_i$. Thus

$$\lim t_i px_o = \lim t_i p\alpha_i x_o \in \cap \overline{U}_i x_o \subseteq \cap N_i = \{vx_o\}.$$

and $$\lim t_i vx_o \in \cap \overline{U}_i x_o \subseteq \cap N_i = \{vx_o\}.$$

It follows that px_o and vx_o are proximal L is a subset of K and K is a residual subset of $\overline{wF}$,

(2) Since $Kx_o = W$, W is dense in $\overline{wF}x_o = \overline{wFx_o}$. On the other hand

$$W_i = \{x \in \overline{wFx_o} \mid (x,vx) \in T(N_i \times N_i)\}$$

is an open subset of $\overline{wFx_0}$ and $\cap W_i = W$. It follows that W is a residual subset of $\overline{wFx_0}$ and the proof is complete. //

7.2. THEOREM: *Let X be a metric minimal flow such that there exists a point $x_0 \in X$ for which the set*

$$P[x_0] = \{x \in X \mid x \text{ is proximal to } x_0\}$$

is countable. Then X is a PI-flow. In particular a point-distal flow is PI.

Proof: Since u is an arbitrary idempotent of M we can assume that $ux_0 = x_0$. Let (Y_∞, y_∞) be the canonical PI-flow associated with (X,x_0) and let

$$(X_\infty, x_\infty) = (X,x_0) \vee (Y_\infty, y_\infty) \xrightarrow{\phi_\infty} (Y_\infty, y_\infty)$$

be the projection map, then ϕ_∞ is RIC and $H(F)A = F$ where

$$F = \mathcal{G}(Y_\infty, y_\infty) \text{ and } A = \mathcal{G}(X,x_0) = \mathcal{G}(X_\infty, x_\infty).$$

Let

$$Q = \{z \in X_\infty \mid \phi(z) = y_\infty \text{ and } z \text{ and } x_\infty \text{ are proximal}\}.$$

Now if $z \in Q$ then $z = (x,y_\infty)$ for some $x \in X$ and since $x_\infty = (x_0, y_\infty)$ it follows that x and x_0 are proximal points of X. Thus by our assumption Q is countable. Let $w \in \overline{F}$ be an idempotent for which $\overline{wF}$ is an F-minimal subset of $\overline{F}$, then clearly $\overline{wFx_\infty}$ is a subset of $\phi_\infty^{-1}(y_\infty)$. Hence

$$W = \{z \in X_\infty \mid z \in \overline{wFx_\infty} \text{ and } z \text{ is proximal to both } x_\infty \text{ and } wx_\infty\} \subset Q$$

is countable. By proposition 7.1.(2) W is the intersection of two residual subsets of $\overline{wFx_\infty}$ (take $v = u$ and $v = w$ in proposition 7.1.) and therefore W is a countable and residual subset of $\overline{wFx_\infty}$.

Let W_0 be the subset of W which consists of points which are open in $\overline{wFx_\infty}$. Since each point of $W \setminus W_o$ is nowhere dense and since $W \setminus W_o$ is countable W_o is residual and therefore dense in $\overline{wFx_\infty}$. Now wFx_∞ is dense in $\overline{wFx_\infty}$ and it follows that every point of W_o contains an element of wFx_o i.e. $W_o \subsetneq wFx_\infty$. By I.3.2.(3) all the points of W_o are distal to wx_∞, but by the definition of W all the points of W_o are proximal to wx_∞ and it follows that W_o is a singleton. Since W_o is dense in $\overline{wFx_\infty}$ we have $wFx_\infty = wx_\infty$ and therefore also $Fx_\infty = x_\infty$ i.e. $F \subset A = \mathcal{G}(X_\infty, x_\infty)$. Thus $A = F$ and since ϕ_∞ is RIC this implies that ϕ_∞ is an isomorphism. By theorem 4.2.(5) X is a PI-flow.

//

7.3. THEOREM: *Let X be a minimal metric flow. If the enveloping semigroup of X contains only finitely many minimal ideals then X is a PI-flow. In particular if the enveloping semigroup of X contains only one minimal ideal or equivalently if the proximal relation on X is an equivalence relation then X is PI.*

Proof: With notations as in the proof of theorem 7.2. put

$$K = \{p \in \overline{wF} \mid px_\infty \text{ and } x_\infty \text{ are proximal.}\}.$$

Then by 7.1.(1). K is a residual subset of $\overline{wF}$. Suppose now that the enveloping semigroup of X contains n minimal ideals $I_1,\ldots,I_n$. Let $M_1,\ldots,M_n$ be n minimal ideals in βT which are mapped onto $I_1,\ldots I_n$ and let $J_1,\ldots,J_n$ be the sets of idempotents in the corresponding I's. By I.3.2.2.

$$P[x_o] = \{x \in X \mid x \text{ is proximal to } x_o\} = \bigcup_{i=1}^{n} J_i x_o.$$

Define

$$K_i = \{p \in \overline{wF} \mid px_o \in J_i x_o\} \qquad i = 1,2,3,\ldots;$$

then we claim that $\bigcup_{i=1}^{\infty} K_i = K$. Indeed, if $p \in \overline{wF}$ then $py_\infty = y_\infty$ so that px_o proximal to x_o implies $px_\infty = p(x_o,y_\infty) = (px_o,y_\infty)$ is proximal to $x_\infty = (x_0, y_\infty)$. This shows that $K_i \subseteq K$; conversely if $p \in K$ then $px_\infty = p(x_0,y_\infty) = (px_0,y_\infty)$ is proximal to $(x_0,y_\infty) = x_\infty$ and hence px_0 is proximal to x_0 which implies that p belongs to some K_i.

Next we claim that if $\alpha \in F\setminus A$ then for each i, $K_i\alpha \cap K_i = \emptyset$. To see that we observe that if p and also q are in K_i and $q = p\alpha$ then there exist $v_1, v_2 \in J_i$ such that $px_0 = v_1x_0$ and $qx_0 = v_2x_0$. This implies that qx_0 is proximal to px_0 (e.g. $v_1(px_0) = v_1(v_1x_0) = v_1x_0 = px_0$ and also $v_1(qx_0) = v_1(v_2x_0) = v_1x_0 = px_0$). On the other hand if $v \in J$ is such that $vp = p$ then also $vq\alpha = vp\alpha = p_\alpha = q$ and by I.3.2.(3) px_0 and qx_0 are also distal. Thus $px_0 = qx_0 = p\alpha x_0$ hence $up^{-1}p\alpha x_0 = \alpha x_0 = x_0$ and $\alpha \in A$ which is a contradiction.

We now show that F/A is finite. If not then there are elements $\alpha_1,\ldots,\alpha_n$ in F such that for every $1 \le i \le k \le n$, $\alpha_i\alpha_{i+1}\cdots\alpha_k$ is not in A.

If we recall now that for $\alpha \in F\setminus A$, $K_i\alpha \cap K_i = \emptyset$ then it is clear that for each $1 \le i \le n$

$$L_i = (\cdots(((K_i\alpha_1 \cap K)\alpha_2 \cap K)\alpha_3 \cap K)\cdots \cap K)\alpha_n \cap K$$

must be empty. On the other hand $\bigcup_{i=1}^{n} = L$ where

$$L = (\cdots(((K\alpha_1 \cap K)\alpha_2 \cap K)\alpha_3 \cap K)\cdots \cap K)\alpha_n \cap K.$$

But since K is residual in $\overline{wF}$ so is L which therefore cannot be empty. This is a contradiction to our assumption that F/A is infinite and we conclude that it is finite; whence so is Fx_∞.

Since ϕ_∞ is RIC this implies:

$$\phi_\infty^{-1}(py_\infty) = p\circ Fx_\infty = pFx_\infty$$

for every $p \in M$, and by I.4.1.(3). ϕ_∞ is distal. Now K is dense in $\overline{wF}$ and $\overline{wF}x_\infty \subseteq \phi_\infty^{-1}(y_\infty)$ hence Kx_∞ is dense in $\overline{wF}x_\infty$ and by the distality of $\phi_\infty Kx_\infty$ and hence also $\overline{wFx_\infty}$ are equal to the singleton $\{x_\infty\}$. It now follows that $wFx_\infty = x_\infty$ hence $Fx_\infty = x_\infty$ and $F = A$. Since ϕ_∞ is RIC this implies that ϕ_∞ is an isomorphism, therefore $X_\infty = Y_\infty$ and by theorem 4.2.(5) X is PI. //

We do not know whether a metric minimal flow with a countably infinite number of minimal ideals in its enveloping semigroup is necessarily PI. However the converse of this statement (and therefore also the converse of theorem 7.3.) does not hold i.e. there exists a minimal metric PI-flow in whose enveloping semigroup there are uncountably many minimal ideals as the example in VIII.2.2. shows.

7.4. COROLLARY: Let T be an abelian group and let X be a non-trivial minimal weakly mixing flow; then the enveloping semigroup of X contains infinitely many minimal ideals.

Proof: By 7.3. X must be PI if its enveloping semigroup contains only a finite number of minimal ideals. Since every minimal flow of an abelian group is incontractible, by 4.4 if X is PI it admits a non-trivial almost periodic factor. Finally it is well known that a weakly mixing minimal flow of an abelian group does not admit a non-trivial almost periodic factors. (See [29].) //

7.5 THEOREM: (Furstenberg) Let X be a metric minimal distal flow. Then $X = X_\infty = Y_\infty$ and in the canonical tower of Y_∞ all the proximal homomorphisms are isomorphisms.

Proof: By theorem 7.2. $X_\infty = Y_\infty$. Thus all we have to show is that in the canonical tower, built in section 4, the proximal extensions ψ_β are isomorphisms. Indeed this follows by induction using the fact that a distal extension is RIC (lemma 1.2.) and proposition 3.2.(3). //

We remark that in order to obtain a similar proof for the Veech-Ellis point distal structure theorem, it suffices to show that when (X,x_o) is a metric point distal flow with distal point x_o, and $(X,x_o) \xrightarrow{\phi} (Y,y_o)$ is a homomorphism then in the diagram constructed in proposition 3.2. the map θ is almost one-to-one (i.e. $\theta^{-1}(y_o)$ is a singleton). This is really the case (see [3]) but the only proof of this fact that we know of, uses the Veech-Ellis theorem.

BIBLIOGRAPHY

(1) Auslander, J., On the proximal relation in topological dynamics, Proceedings of the American Mathematical Society 11(1960), 890-895.

(2) Auslander, J., Regular minimal sets, I, Transactions of the American Mathematical Society 123(1966), 469-479.

(3) Auslander, J. and Glasner, S., Distal and highly proximal extensions, (to appear).

(4) Auslander, L., Green, L., and Hahn, F., Flows on Homogeneous Spaces, Princeton University Press, Princton, 1963, Annals of Mathematics Studies 53.

(5) Azencott, R., Espace de Poisson des groupes localement compacts, Berlin, Springer-Verlag, 1970 (Lecture Notes in Mathematics 148).

(6) Choquet, G. and Denz, J., Sur l'equation de convolution $\mu = \mu * \sigma$, C. R. Acad. Sc., Paris 250(1960), 799-801.

(7) Dunford, N. and Schwartz, J., Linear Operators, vol. 1, Interscience Publichers, New York, 1958.

(8) Ellis, R., Locally compact transformation groups, Duke Mathematical Journal 24(1957), 119-125.

(9) Ellis, R., Lectures on Topological Dynamics, W. A. Benjamin, New York, 1969.

(10) Ellis, R., The Veech structure theorem, Transactions of the American Mathematical Society 186(1973), 203-218.

(11) Ellis, R. and Keynes, H., Bohr compactifications and a result of Folner, Israel Journal of Mathematics, 12(1972), 314-330.

(12) Ellis, R., Glasner, S., and Shapiro, L., P I-Flows, Advances in Mathematics, (to appear).

(13) Furstenberg, H., The structure of distal flows, American Journal of Mathematics 85(1963), 477-515.

(14) Furstenberg, H., A Poisson formula for semi-simple Lie groups, Annals of Mathematics 77(1963), 335-383.

(15) Furstenberg, H., Translation invariant cones of functions on semi-simple Lie groups, Bulletin of the American Mathematical Society 71(1965), 271-326.

(16) Furstenberg, H., Disjointness in ergodic theory, minimal sets, and a problem in Diophantine approximation, Mathematical Systems Theory 1(1967), 1-49.

(17) Furstenberg, H., Random walks and discrete subgroups of Lie groups, Advances in Probability and Relatent Topics, Vol. 1, Dekkers, New York 1911, 1-63.

(18) Furstenberg, H., Boundaries of Riemannian symmetric spaces, Symmetric Spaces, Short courses presented at Washington University, New York 1972.

(19) Furstenberg, H., Boundary theory and stochastic processes on homogeneous spaces, Harmonic Analysis on Homogeneous Spaces, Symposia in Pure Mathematics, Williamstown, Mass., 1972.

(20) Glasner, S., Topological dynamics and group theory, Transactions of the American Mathematical Society 187(1974), 327-334.

(21) Glasner, S., Compressibility properties in topological dynamics, American Journal of Mathematics 97(1975), 148-171.

(22) Gottschalk, W. and Hedlund, G., Topological Dynamics, American Mathematical Society Colloquium Publications vol. 36, Providence, 1955.

(23) Greenleaf, F., Invariant Means on Topological Groups, Van Nostrand Reinhold Company, New York, 1969.

(24) Guivarch, Y., Croissance polynomials et periodes den fonctions harmoniques, Bulletin de la Societe Mathematique de France 101(1973), 333-379.

(25) Hahn, F., A fixed point theorem, Mathematical Systems Theory 1(1967), 55-57.

(26) Helgason, S., Differential Geometry and Symmetric Spaces, Academic Press, New York, 1962.

(27) Hewitt, E. and Ross, K., Abstract Harmonic Analysis, Springer-Verlag, Berlin, 1963.

(28) Jenkins, J., Growth of connected locally compact groups, Journal of Functional Analysis 12(1973), 113-127.

(29) Keynes, H. and Robertson, J., Eigenvalue theorems in topological transformation groups, Transactions of the American Mathematical Society 139(1969), 359-369.

(30) Knapp, A., Functions behaving like almost automorphic functions, Topological Dynamics - an International Symposium, W. A. Benjamin, New York, 1968, 299-317.

(31) Markley, N., Transitive homeomorphisms of the cinch, Mathematical Systems Theory 2(1968), 247-249.

(32) Milnor, J., Growth of finitely generated solvable groups, Journal of Differential Geometry 2(1968), 447-449.

(33) Moore, C., Compactifications of symmetric spaces, American Journal of Mathematics 86(1964), 201-218.

(34) Moore, C., Compactification of symmetric spaces II: The Cantan domains, American Journal of Mathematics 86(1964), 358-378.

(35) Moore, C., Distal affine transformation groups, American Journal of Mathematics 90(1968), 733-751.

(36) Mostow, Some new decomposition theorems, Memoirs of the American Mathematical Society 14, 31-54.

(37) Peleg, R., Weak disjointness of transformation groups, Proceedings of the American Mathematical Society 33(1972), 165-170.

(38) Phelps, R., Lectures on Choquet's Theorem, Van Nostrand, Princeton, 1966.

(39) Shapiro, L., Proximality in minimal transformation groups, Proceedings of the American Mathematical Society 26(1970), 521-525.

(40) Tits, J., Free subgroups in linear groups, Journal of Algebra 20(1972), 250-270.

(41) Veech, W., Point distal flows, American Journal of Mathematics, 92(1970), 205-242.

(42) Warner, G., Harmonic Analysis on Semi-simple Lie Groups, vol. I, Springer-Verlag, Berlin, 1972.

(43) Weil, A., L'intigration dans les groupes topologiques et ses applications, Hermann, Paris, 1940.

(44) Wolf, J., Growth of finitely generated solvable groups and curvature of Riemannian manifolds, Journal of Differential Geometry 2(1968), 421-446.